Ernst Schering Foundation Symposium
Proceedings 2007-1
Progestins and the Mammary Gland

Ernst Schering Foundation Symposium
Proceedings 2007-1

Progestins
and the Mammary Gland

From Basic Science to Clinical Applications

O. Conneely, C. Otto
Editors

With 43 Figures

 Springer

Series Editors: G. Stock and M. Lessl

Library of Congress Control Number: 2007943074

ISSN 0947-6075
ISBN 978-3-540-73492-5 Springer Berlin Heidelberg New York

Springer is a part of Springer Science+Business Media
springer.com

Cover design: design & production, Heidelberg
Typesetting and production: LE-TEX Jelonek, Schmidt & Vöckler GbR, Leipzig
21/3180/YL – 5 4 3 2 1 0   Printed on acid-free paper

Preface

Steroid hormone receptors are important drug targets and have been the focus of basic and applied research for decades. Steroid hormone receptors act as ligand-dependent transcription factors. Upon ligand binding, the receptors bind to hormone responsive *cis*-acting DNA elements (HREs) in the nucleus and regulate the expression of target genes by recruiting chromatin-modifying activities that either promote or deny access to the basal transcription machinery. In general, agonist ligands recruit coactivator proteins that promote transcriptional activation, while receptor antagonists recruit corepressors that prevent transcriptional activation. The ability of steroid hormone receptors to regulate distinct gene expression profiles in different tissues has been exploited in recent years in the clinical development of novel hormone receptor modulators that have the capability of harnessing the beneficial properties of steroids while eliminating their potential adverse effects. Elucidation of the molecular mechanisms by which steroid receptors elicit distinct transcriptional responses in different tissues is critical to the development of optimal tissue-selective receptor modulators. Recent progress in our understanding of these mechanisms reveals that several levels of complexity may explain the tissue specificity of hormone action. These include distinct tissue-selective expression of receptor isoforms in steroid target tissues, variations in sequence composition of HREs that influence receptor conformation and coregulator recruitment at responsive target genes, different receptor coregulator expression profiles in target tissues, and different cellular signalling contexts.

The progesterone receptor (PR), a member of the nuclear hormone receptor family, is critically involved in mammalian reproduction and mammary gland development. Synthetic progestins are widely used in combined oral contraception (ovulation inhibition) and hormone therapy (inhibition of estradiol-induced uterine epithelial cell proliferation). One potential side effect of progestin action in combined hormone therapy is enhanced proliferation of normal as well as malignant mammary epithelial cells. While clinical trials using the synthetic progestin, medroxyprogesterone acetate, indicate that progestins used in combined hormone therapy may contribute to breast cancer risk (WHI study), the mechanisms by which progestins regulate proliferation of mammary epithelial cells remain poorly understood.

To further our understanding of progestin action in both mammary gland physiology and pathology, and to foster the interaction between basic research and drug development, the Ernst Schering Foundation held a symposium on 'Progestins and the Mammary Gland—From Basic Science to Clinical Applications'. The present volume covers the different areas of progestin research that were the focus of the symposium. Robert Clarke summarized the role of adult tissue stem cells in normal mammary gland development and formation of breast carcinomas and highlighted the role of Wnt signalling downstream of PR activation in these processes. Bert O'Malley discussed the central role of coactivators in mediating distinct tissue-specific transcriptional responses to hormone and introduced the novel concept of the 'ubiquitin clock' that explained how cycles of posttranslational modifications of coactivators via phosphorylation and subsequent ubiquitinylation can turn on and off PR-mediated signalling. The molecular mechanisms of pregnancy-induced mammary gland remodelling were addressed by Orla Conneely. She put emphasis on the important interplay of PR and the prolactin receptor. Using genetically modified mice, she could demonstrate that the PRB isoform is more potent in promoting ductal proliferation and sidebranching than PR-A. Gene expression analysis in the mammary glands of PR-deficient and wild-type mice allowed the identification of paracrine pathways involved in epithelial cell proliferation and morphogenesis. John Lydon developed an elegant genetic mouse model leading to the ablation of the coactivator SRC-2 in all PR-expressing cells of the organism. He provided in vivo evidence for a critical role of the SRC-2 coactivator in mediating tissue selectivity of progesterone action in both the uterus and mammary gland. Using clinical studies as well as gene expression analysis in breast cancer cell culture, Christine Clarke discussed the emergence of aberrant PR isoform expression patterns in human breast cancers that may contribute to deregulated expression of progesterone responsive target genes resulting in changes in morphology, cell adhesion, and invasive behavior. Daniel Medina elaborated on the concept of short-term hormonal exposure to prevent breast cancer that was based on epidemiological observations and animal models. The utility of mathematical models to predict breast cancer risk after hormone therapy was described by Malcolm Pike. Christiane Otto described an approach that exploited nongenomic versus genomic PR-mediated signalling to identify progestins with reduced proliferative activity in the mammary gland. Matt Yudt reported on unexpected findings with a nonsteroidal PR modulator that, depending on context, concentration, and species, behaved as an agonist or antagonist, respectively. Such tool compounds might be very useful for further analysis of species-specific receptor conformations and receptor/coactivator interactions.

Taken together, during the last years, our mechanistic understanding of tissue-specific progestin action has greatly advanced but is still far from being complete. One important take-home message derived from the final discussion of this Ernst Schering Foundation symposium was that antiprogestins should be developed for the treatment of breast cancer.

Orla M. Conneely
Christiane Otto

Contents

List of Editors and Contributors

Editors

Conneely, O.
Department of Molecular and Cellular Biology,
Baylor College of Medicine,
One Baylor Plaza, Houston, Texas 77030, USA
(e-mail: orlac@bcm.tmc.edu)

Otto, C.
TRG Women's Healthcare, Bayer Schering Pharma AG,
13342 Berlin, Germany
(e-mail: christiane.otto@bayerhealthcare.com)

Contributors

Allred, D.C.
Department of Pathology and Immunology,
Washington University School of Medicine,
4550 Scott Avenue, St. Louis, MO 63110 USA

Altmann, H.
TRG Women's Healthcare, Bayer Schering Pharma AG,
13342 Berlin, Germany

Amato, P.
Department of Obstetrics and Gynecology, Baylor College of Medicine,
One Baylor Plaza, Houston, Texas 77054, USA

Arnett-Mansfield, R.
Department of Molecular and Cellular Biology,
Baylor College of Medicine,
One Baylor Plaza, Houston, Texas 77030, USA

Berrodin, T.J.
Womens's Health and Musculoskeletal Biology, Wyeth Research,
500 Arcola Road, Collegeville, PA 19426, USA

Bretz, J.
Womens's Health and Musculoskeletal Biology, Wyeth Research,
500 Arcola Road, Collegeville, PA 19426, USA

Chippari, S.
Womens's Health and Musculoskeletal Biology, Wyeth Research,
500 Arcola Road, Collegeville, PA 19426, USA

Clarke, C.L.
Westmead Institute for Cancer Research, University of Sydney
at Westmead Millennium Institute Department
of Translational Oncology, Westmead Hospital,
Darcy Rd, Westmead, NSW 2145, Australia
(e-mail: christine_clarke@wmi.usyd.edu.au)

Clarke, R.B.
Breast Biology Group, Cancer Studies, University of Manchester,
Paterson Insitute for Cancer Research,
Wilmslow Road, Manchester, M20, 4BX, UK
(e-mail: robert.clarke@manchester.ac.uk)

Cohen, J.
Womens's Health and Musculoskeletal Biology, Wyeth Research,
500 Arcola Road, Collegeville, PA 19426, USA

Fensome, A.
Chemical and Screening Sciences Wyeth Research, 500 Arcola Road,
Collegeville, PA 19426, USA

Fritzemeier, K.-H.
TRG Women's Healthcare, Bayer Schering Pharma AG,
13342 Berlin, Germany

Fuchs, I.
TRG Women's Healthcare, Bayer Schering Pharma AG,
13342 Berlin, Germany

Fuqua, S.A.W
Department of Molecular and Cellular Biology, and Baylor Brest Center,
Baylor College of Medicine,
One Baylor Plaza, Houston, Texas 77030, USA

Graham, J.D.
Westmead Institute for Cancer Research, University of Sydney
at Westmead Millennium Institute, Translational Oncology,
Westmead Hospital, Darcy Rd, Westmead, NSW 2145, Australia

Han, S.J.
Department of Molecular and Cellular Biology,
Baylor College of Medicine,
One Baylor Plaza, Houston Texas 77030, USA

Harrison, H.
Breast Biology Group, Cancer Studies, University of Manchester,
Paterson Institute for Cancer Research,
Wilmslow Road, Manchester, M20, 4BX, UK

Kittrell, F.S.
Department of Molecular and Cellular Biology, and Baylor Brest Center,
Baylor College of Medicine,
One Baylor Plaza, Houston, Texas 77030, USA

Klewer, M.
TRG Women's Healthcare, Bayer Schering Pharma AG,
13342 Berlin, Germany

Lamb, R.
Breast Biology Group, Cancer Studies, University of Manchester,
Paterson Institute for Cancer Research,
Wilmslow Road, Manchester, M20, 4BX, UK

Lee, S.
Department of Preventive Medicine,
Norris Comprehensive Cancer Center,
University of Southern California,
1441 Eastlake Avenue, Los Angeles, California 90033, USA

Lundeen, S.
Women's Health and Musculoskeltal Biology, Wyeth Research,
500 Arcola Road, Collegeville, PA 19426, USA

Lydon, J.P.
Department of Molecular and Cellular Biology,
Baylor College of Medicine,
One Baylor Plaza, Houston Texas 77030, USA
(e-mail: jlydon@bcm.tmc.edu)

DeMayo, F.J.
Department of Molecular and Cellular Biology,
Baylor College of Medicine,
One Baylor Plaza, Houston Texas 77030, USA

Mote, P.A.
Westmead Institute for Cancer Research, University of Sydney
at Westmead Millennium Institute, Translational Oncology,
Westmead Hospital, Darcy Rd, Westmead, NSW 2145, Australia

Mulac-Jericevic, B.
Department of Molecular and Cellular Biology,
Baylor College of Medicine,
One Baylor Plaza, Houston, Texas 77030, USA

O'Malley, B.W.
Department of Molecular and Cellular Biology,
Baylor College of Medicine,
One Baylor Plaza, Houston Texas 77030, USA

Medina, D.
Department of Molecular and Cellular Biology, and Baylor Brest Center,
Baylor College of Medicine,
One Baylor Plaza, Houston, Texas 77030, USA
(e-mail: dmedina@bcm.tmc.edu)

Mukherjee, A.
Department of Molecular and Cellular Biology,
Baylor College of Medicine,
One Baylor Plaza, Houston, Texas 77054, USA

Pearce, C.L.
Department of Preventive Medicine,
Norris Comprehensive Cancer Center,
University of Southern California,
1441 Eastlake Avenue, Los Angeles, California 90033, USA

Pike, M.C.
Department of Preventive Medicine,
Norris Comprehensive Cancer Center,
University of Southern California,
1441 Eastlake Avenue, Los Angeles, California 90033, USA
(e-mail: mcpike@usc.edu)

Rohde-Schulz, B.
TRG Women's Healthcare, Bayer Schering Pharma AG,
13342 Berlin, Germany

Slayden, O.
Oregon National Primate Center, Portland, OR 97201, USA

Spicer, D.V.
Department of Preventive Medicine,
Norris Comprehensive Cancer Center,
University of Southern California,
1441 Eastlake Avenue, Los Angeles, California 90033, USA

Schwarz, G.
TRG Women's Healthcare, Bayer Schering Pharma AG,
13342 Berlin, Germany

Tsimelzon, A.
Department of Molecular and Cellular Biology, and Baylor Brest Center,
Baylor College of Medicine,
One Baylor Plaza, Houston, Texas 77030, USA

Winneker, R.C.
Womens's Health and Musculoskeletal Biology, Wyeth Research,
500 Arcola Road, Collegeville, PA 19426, USA

Wrobel, J.
Woman's Health and Musculoskeletal Biology, Wyeth Research,
500 Areda Road, Collegeville, PA 19426, USA
(e-mail: wrobelj@wyeth.com)

Wu, A.H.
Department of Preventive Medicine,
Norris Comprehensive Cancer Center,
University of Southern California,
1441 Eastlake Avenue, Los Angeles, California 90033, USA

Yudt, M.R.
Womens's Health and Musculoskeletal Biology, Wyeth Research,
500 Arcola Road, Collegeville, PA 19426, USA
(e-mail: yudtm@wyeth.com)

Zhang, P.
Chemical And Screening Sciences Wyeth Research,
500 Arcola Road, Collegeville, PA 19426, USA

Zhang, Z.
Womens's Health and Musuloskeletal Biology, Wyeth Research,
500 Arcola Road, Collegeville, PA 19426, USA

Zhu, Y.
Womens's Health and Musculoskeletal Biology, Wyeth Research,
500 Arcola Road, Collegeville, PA 19426, USA

Ernst Schering Foundation Symposium Proceedings, Vol. 1, pp. 1–23
DOI 10.1007/2789_2008_074
© Springer-Verlag Berlin Heidelberg
Published Online: 29 February 2008

Mammary Development, Carcinomas and Progesterone: Role of Wnt Signalling

R. Lamb, H. Harrison, R.B. Clarke^(✉)

Breast Biology Group, Cancer Studies, University of Manchester, Paterson Institute for Cancer Research, Wilmslow Road, M20 4BX Manchester, UK
email: *robert.clarke@manchester.ac.uk*

Abstract. The mammary gland begins development during embryogenesis but after exposure to hormonal changes during puberty and pregnancy undergoes extensive further development. Hormonal changes are key regulators in the cycles of proliferation, differentiation, apoptosis and remodelling associated with pregnancy, lactation and involution following weaning. These developmental processes within the breast epithelium can be explained by the presence of a long-lived population of tissue-specific stem cells. The longevity of these stem cells makes them susceptible to accumulating genetic change and consequent transformation. The ovarian steroid progesterone, acting via the secreted factor Wnt4, is known to be essential for side branching of the mammary gland. One function of Wnt proteins is self-renewal of adult tissue stem cells, suggesting that progesterone may exert its effects within the breast, at least partly, by regulating the mammary stem cell population.

1   Introduction

This review aims to discuss the role of progesterone and its downstream targets such as Wnt in mammary gland development and breast carcinomas. Evidence is accumulating to suggest that stem cells (SCs) are involved in both normal mammary gland development and the formation of breast carcinomas. We hypothesise that progesterone may have a previously unrecognised role of signalling through the Wnt pathway to increase SC self-renewal.

2   Mammary Gland Development

Mammary gland development begins during embryogenesis, with the formation of a rudimentary ductal system and remains virtually unaltered throughout childhood (Naccarato et al. 2000). During puberty, hormonal changes induce the formation of networks of epithelial ducts which grow outwards from the nipple and divide into primary and secondary ducts ending with bud structures. From these end buds and branching ductal system, terminal ductal lobuloalveolar units (TDLUs) or lobules form that are the functional milk-producing glands of the pre-menopausal breast. Each lobule is lined by a layer of luminal epithelial cells surrounded by a basal layer of myoepithelial cells. The TDLU is the site from which many epithelial hyperplasias and carcinomas of the breast are thought to arise (Wellings et al. 1975). Full development of mammary gland occurs during pregnancy with accelerated growth of the TDLUs in preparation for lactation (Hovey et al. 2002). At weaning, massive involution and remodelling of the tissue occurs returning the gland to its non-pregnant state (Furth et al. 1997). This cycle of pregnancy-associated proliferation, differentiation, apoptosis and remodelling can occur many times during the reproductive lifespan of mammals and depends on a long-lived population of tissue-specific SCs that have a near infinite propensity to produce functional cells.

3   Role of Progesterone in Mammary Gland Development

Ovarian steroids play a key role in the proliferation and differentiation of mammary epithelium. In terms of biological activity, oestradiol (E2) and progesterone (P) are, respectively, the most important oestrogen and progestogen circulating in women. From the onset of menarche until menopause, these hormones, in the absence of pregnancy, are synthesised in a cyclical manner.

A wealth of data exists which provides evidence that both E2 and P are important in mammary gland development and tumour formation. Clinical management of females with gonadal dysgenesis or gonadotrophin insufficiency shows that E2 is necessary but not sufficient to induce puberty and breast development (Laron et al. 1989). Additionally, reduced levels of exposure to E2 and P with either artificially induced or naturally early menopause significantly reduce the risk of developing breast cancer. Conversely, increased exposure through early menarche, late menopause, or late age at first full term pregnancy all raise the risk of developing breast cancer. This increased risk is also observed with the use of exogenous ovarian hormones in the form of the oral contraceptive pill or hormone replacement therapy (Clemmons and Gross 2001; Travis and Key 2003).

Ovarian hormones have been shown to exert their effects through ligand-activated steroid receptors in the mammary epithelium. Approximately 10%–15% of the cells within the epithelium coexpress oestrogen receptor alpha (ERα) and progesterone receptor (PR) and are known to be located in the luminal epithelia of the ductal and lobular structures (Clarke et al. 1997; Petersen et al. 1987). A recent study showed that the ERα homozygous knockout mouse model showed no mammary development beyond the formation of the rudimentary structure at embryogenesis (Mallepell et al. 2006). However, when cells from the ERα knockout mouse (ER$\alpha^{-/-}$) are mixed with wild-type ERα cells before they are engrafted into the cleared fat pad of a recipient mouse, ER$\alpha^{-/-}$ cells are able to proliferate and contribute to normal mammary gland development, suggesting that E2 elicits secretion of local factors (Mallepell et al. 2006). During development, oestrogen is responsible for ductal elongation whereas P is responsible for side branching. The PR$^{-/-}$ mouse shows that PR is essential for ductal side branching and the alveolar development of the mammary gland whereas chimeric epithelia of PR$^{-/-}$ cells and wild-type cells undergo complete alveolar development, suggesting a secreted local factor (Brisken et al. 1998). Together these data suggest that the proliferation of ERα/PR-negative cells is controlled by a paracrine mediator of the systemic hormonal signal and fits with the finding that proliferating cells are ERα/PR-negative in the mouse, rat and human mammary epithelium (Clarke et al. 1997; Russo et al. 1999; Seagroves et al. 1998).

Loss of ductal side branching of the mammary epithelium of PR$^{-/-}$ mice can be rescued by the ectopic expression of Wnt1. The Wnt pathway is therefore likely to be downstream of P signalling and acts in a paracrine manner (Brisken et al. 2000). Wnt1 is not expressed in normal human mammary epithelium; however, the closely related protein Wnt4 is expressed during the period when side branching occurs in early pregnancy in the mouse (Gavin and McMahon 1992; Weber-Hall et al. 1994). Although Wnt4$^{-/-}$ mice die during embryonic development, transplantation of murine mammary epithelium from Wnt4$^{-/-}$ embryos showed that Wnt4 had an essential role in ductal side branching in early pregnancy. Furthermore, this study showed that P induces the expression of Wnt4 mRNA, which co-localises with PR in the luminal compartment of the ductal epithelium (Brisken et al. 2000). In a recent investigation, P was found to be essential for priming ductal cells to form side branches and alveoli in response to Wnts, suggesting a further level of complexity in signalling (Hiremath et al. 2007). Cumulatively, this evidence suggests that Wnt signalling is essential for mediating P function during mammary gland development.

4   Adult Stem Cells

Adult SCs are a small pool of tissue-specific, long-lived cells that last throughout life and can be defined by their ability to self-renew and to produce differentiated, functional cells within an organ (Dexter and

Spooncer 1987; Jones 1997; Orkin 2000; Watt 1998). Adult SCs are necessary for tissue development, replacement and repair (Fuchs and Segre 2000).

The first tissue-specific adult SCs to be well defined were identified within the bone marrow and termed haematopoietic stem cells (HSCs) (Siminovitch et al. 1963). Transplantation of retroviral-tagged, individual bone marrow cells into a lethally irradiated mouse showed that HSCs were multipotent, having the ability of multi-lineage differentiation generating precursor cells that can differentiate into all mature blood cells (Bonnet 2003; Jordan and Lemischka 1990). Since the discovery of the HSCs, SCs within many other tissues have been identified including the mammary gland. Although HSCs are currently the best characterised, great efforts have been made to further characterise other tissue-specific SCs.

SCs have the ability to undergo either symmetrical or asymmetrical division, depending upon the cellular context. During normal tissue homeostasis, asymmetrical SC division occurs and results in one new SC and one more differentiated daughter cell which will then go on to generate cells which will undergo terminal differentiation down specific cell lineages. This process of SC replacement by a daughter cell is termed self-renewal. Symmetrical division results in the production of either two undifferentiated SCs by self-renewal or two differentiated daughter cells where the SC is lost. It is possible that the first scenario would be required during development and the second would occur during tissue ageing. The subtle balance between symmetrical or asymmetrical division of the SCs is tightly regulated by local factors to restrict the number of SCs during normal tissue homeostasis and increase the population of SCs during tissue development and repair (Potten and Loeffler 1990).

5   Mammary Epithelial Stem Cells

Cyclic proliferation, differentiation, apoptosis and remodelling of the mammary gland suggests the presence of a long-lived population of tissue-specific SCs. Unlike differentiated cells, which have a relatively short life span, SCs' longevity makes them susceptible to accumulating genetic damage and they represent likely targets for carcinogenic transformation. As a consequence, cancer may be a SC disease, suggesting that successful breast cancer prevention strategies must be targeted to mammary epithelial SCs.

The first evidence to support the notion of mammary SCs came from murine transplantation experiments. Mammary gland tissue was removed from a donor mouse and transplanted into the cleared mammary fat pad of a recipient mouse, regenerating a fully functional mammary gland (Deome et al. 1959). More recently, transplantation of mammary epithelia marked with mouse mammary tumour virus (MMTV) showed that single epithelial cell clones were capable of regenerating a complete, lactationally functional ductal and alveolar system after transplantation into cleared mammary fat pads. Serial transplantation of these cells was able to recapitulate the mammary gland, demonstrating self-renewing and multipotent characteristics of the cells (Kordon and Smith 1998).

6   Side Population Analysis

Recently, several methods have been used to identify the mammary epithelium SCs or stem-like cells. One such method is side population analysis, which has previously been used to identify HSCs (Goodell et al. 1997). Studies within the mouse showed that a sub-population of mammary epithelial cells defined by its ability to efflux the dye Hoechst 33342 and termed the "side population" (SP) includes these transplantable mammary SCs. In addition, these cells represent approximately 2–3% of epithelial cells and are enriched for putative SC markers such as Sca1 and α6-integrin (Alvi et al. 2003; Liu et al. 2004; Welm et al. 2002). This method has also been used to analyse the SP within normal human breast tissue, showing comparable results to those observed in the mouse. The percentage of cells which were able to efflux the dye and form the SP varied from 0.2% to 1% to 5% (Alvi et al. 2003; Clarke et al. 2005; Clayton et al. 2004; Dontu et al. 2003; Liu et al. 2004; Welm et al. 2002). The differences between these frequencies can be accounted for by the variation in methodologies used by different groups. Colony growth from single cells in non-adherent culture systems has also been used to identify SCs which were pioneered for the growth of neurospheres from brain tissue, which were enriched for neural SCs (Dontu et al. 2003). Using this technique human breast cells grow "mammospheres" of which 27% of the total sphere cells were found to be within the SP. Additionally, from fresh breast digests, only SP cells and not non-SP cells were able to form mammospheres (Dontu et al. 2003). Despite these encouraging results, there are a number of issues which must be considered with this technique. SP and non-SP cells do not form completely discrete groups. Freshly isolated cells from murine mammary tissue, when transplanted into the cleared fat pad of a recipient mouse, were able to form functional mammary glands. This observation was not limited to the SP cells (5/37 outgrowths), since non-SP cells (6/25 outgrowths) were also able to produce an outgrowth (Alvi et al. 2003). These data suggest that SP analyses are not directly isolating the mammary SCs; they may be missing some cells that do not have the ability to efflux the dye. It is also possible that the dye Hoechst 33342 is toxic to cells; perhaps cells that can efflux the dye are able to form mammary glands and mammospheres simply because they are left unharmed when compared to the cells which are unable to efflux the dye. This method, therefore, may not be the most suitable for the identification of mammary SCs (Smalley and Clarke 2005).

7 Cell Surface Markers

A more appropriate method may be the analysis of cell surface markers that should avoid harm to the cells and any affect on downstream assays of SC potential. A number of studies using cell surface markers have been carried out in both mice and humans. One study showed that a single mouse mammary cell from a subpopulation that was negative for known lineage markers (Lin$^-$) and positive for the cell surface markers CD29 and CD24 (Lin$^-$/CD29hi/CD24$^+$) was able to reconstitute the cleared mammary fat pad of a recipient mouse. Only 1/64 cells from this subpopulation had the ability to produce the normal heterogeneous structure of the gland, suggesting that these cell markers are not sufficient to completely mark the SC (Shackleton et al. 2006). The subpopulation can be further enriched using a CD49f$^+$ sort with 1 in 20 mouse mammary cells from this population having the ability to regenerate the entire gland. There was also evidence of up to 10 symmetrical self-divisions (Stingl et al. 2006). In another study, primary mouse mammary epithelial cells sorted for the cell surface markers CD49f[+], CD24[+], endoglin[+] and PrP[Med] showed the greatest propensity to generate mammospheres (floating colonies) in non-adherent suspension culture in vitro. Furthermore, mammospheres were able to regenerate the entire mammary gland upon transplantation into a mouse mammary fat pad (Liao et al. 2007).

A ductally located SC zone has been identified in the human breast where cells were observed to express the SC proteins SSEA-4, keratin 5 (K5), K6a, K15 and Bcl-2. These cells were shown to be quiescent, like some other adult tissue SCs, and surrounded by basement membrane rich in chondroitin sulphate. Colony formation and mammosphere formation assays provide evidence that these ductal cells have SC properties (Villadsen et al. 2007). In contrast, the progenitors were observed to have a higher rate of proliferation, were found outside of the ductal zone and were surrounded by basement membrane rich in laminin-2/4.

8   Cancer Stem Cells

Carcinomas are believed to arise through a series of mutations that may occur over many years. SCs, by their long-lived nature, are exposed to damaging agents for long periods of time. Accumulation of mutations within these cells could result in their transformation, and consequently mammary SCs may be the source of mammary carcinomas. Alternatively, mutations, or at least the final transforming mutation, could arise during transit amplification of progenitor cells and lead to acquisition of self-renewal ability. Either of the above scenarios could theoretically generate cancer SCs (CSCs) which act to generate the tumour, and are the tumorigenic or cancer-initiating cells that form a minor subpopulation within each breast tumour, necessary for its propagation (Al-Hajj and Clarke 2004).

Current treatments such as radiotherapy target the main proliferating mass of the tumour leaving the source of the cancer, the CSCs (or cancer-initiating cells) unaffected (Chen et al. 2007; Phillips et al. 2006;

Woodward et al. 2007). Consequently, the CSCs survive and are likely to be responsible for recurrence of the carcinomas. It is therefore essential to develop therapies that target the CSC itself. To aid the development of such treatments, the breast CSC must first be identified and isolated. Multiple investigations are currently on-going to address this issue to identify and study human mammary SCs.

Using a model in which human breast cancer cells were grown in immunocompromised mice, as few as 100 cells with the cell surface markers CD44$^+$CD24$^{-/low}$ from 8/9 patient samples were tumourigenic in mice. CD24 is expressed on more differentiated cells whereas CD44 is expressed on more progenitor-like cells. The tumourigenic population of cells marked by the CD44$^+$CD24$^{-/low}$ lineage could be serially passaged to generate new tumours (Al-Hajj et al. 2003). Since this initial study, many others have attempted to link CD44 and CD24 with mammary SCs. Cultured cells from human breast cancer lesions marked by CD44$^+$CD24$^{-/low}$ lineage were capable of (1) self-renewal, (2) extensive proliferation as clonal non-adherent spherical clusters termed mammospheres and (3) differentiation along different mammary epithelial lineages. Furthermore, as few as 10^3 of these cells were required to induce tumour formation in the mammary fat pad of severe combined immunodeficiency (SCID) mice. The mammosphere formation assay is a suitable in vitro model to study breast cancer initiating cells and potential therapeutic targets (Ponti et al. 2005).

Studies of CD44 and CD24 expression in primary breast tumours indicate that expression correlates with patient survival and CD44$^+$ CD24$^{-/low}$ cells from breast cancer cell lines appear to be more invasive (Sheridan et al. 2006; Shipitsin et al. 2007). In addition, breast cells which express the putative "SC marker" CD44$^+$CD24$^{-/low}$ phenotype express genes involved in cell motility and angiogenesis. Phenotypically, these cells are more mesenchymal, motile and are predominately oestrogen receptor negative (Shipitsin et al. 2007). This observation has also been observed in other cell lines with a basal-like/mesenchymal phenotype that have also been reported to have an increased sub-population of CD44$^+$CD24$^{-/low}$ cells (Sheridan et al. 2006). Interestingly, cell lines which are more phenotypically luminal epithelial express less CD44 and more CD24. It has been suggested recently that CD44$^+$ cells are predominately basal-like and therefore are present in poor prognosis basal-like tumours, whereas CD24$^+$ cells are luminal-like and are predominant in more differentiated luminal-type cancers (Fillmore and Kuperwasser 2007).

Analysis of human mammary SCs has proved more difficult compared to analysis in micebecause of the limitations of experimentation with humans. Humanisation of the mouse mammary fat pad, however, is possible using co-transplantation of human stromal fibroblasts, permitting both normal and tumour cells to be implanted and to reconstruct human breast tissue using a mouse model (Kuperwasser et al. 2004; Proia and Kuperwasser 2006).

It will be vital in the future to identify additional cell markers to further enrich for the tumourigenic cell population and eventually to obtain a pure population of SCs. Once this cell population can be identified then regulatory pathways determining the SC phenotype, SC self-renewal and survival can be discovered. This will help the development of drugs that specifically target cancer SCs with the hope that these drugs will eradicate the SCs at the root of the cancer, prevent recurrence and improve mortality.

9 The Wnt Pathway

A number of regulatory signalling pathways are reported to be involved in normal mammary SCs including the Hedgehog, Notch, leukaemia inhibitory factor (LIF), transforming growth facto-β(TGF-β) and epidermal growth factor (EGF) families, prolactin, oestrogen and P, and Wnt (Boulanger et al. 2005; Clarke et al. 2005; Dontu et al. 2003, 2004; Dontu and Wicha 2005; Ewan et al. 2005; Kritikou et al. 2003; Li et al. 2003; Liu et al. 2004; Fig. 1). These pathways are known to be dysregulated in many cancers, including the breast. (Chang et al. 2007; Clarke et al. 2004; Hatsell and Frost 2007; Hu et al. 2004; Johnston et al. 2006; Stylianou et al. 2006; Turashvili et al. 2006; Tworoger and Hankinson 2006).

The Wnt pathway is of particular interest as it is downstream of P (Fig. 2). Wnts are a family of secreted, cysteine-rich glycoproteins associated with the extracellular matrix and the cell surface (Parkin et al. 1993; Schryver et al. 1996). Canonical Wnt signalling is a well-charac-

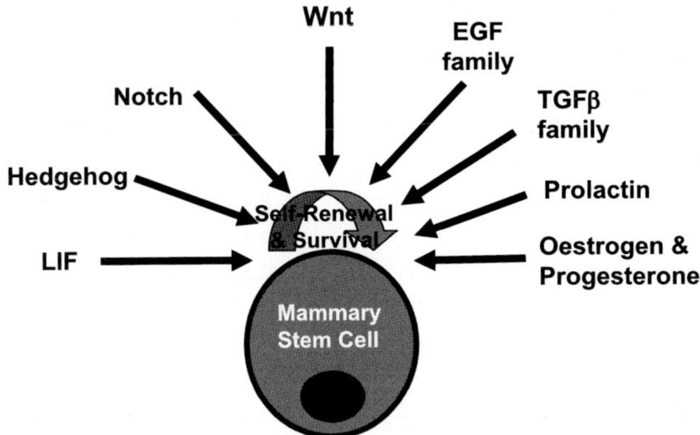

Fig. 1. Mammary stem cell self-renewal signalling pathways. The signalling pathway families Hedgehog, Notch, LIF, TGF-β and EGF, along with oestrogen, progesterone, and Wnt, influence the self-renewal and survival of mammary stem cells. Dysregulation of these pathways is likely to play a role in mammary carcinomas

terised pathway involved in cell–cell adhesion and cell cycle control. Autocrine and paracrine secretion of extracellular Wnt ligands controls the activation of the pathway by binding to the transmembrane family of Frizzled receptors and low-density lipoprotein receptor protein 5/6 (LRP5/6) (Bejsovec 2005; Bhanot et al. 1996). Binding of Wnt to Frizzled receptor results in the phosphorylation of the cytoplasmic mediator Dishevelled, and the inhibition of the multifunctional serine/ threonine kinase GSK3β (Doble and Woodgett 2003). When GSK3β is inactive, β-catenin accumulates within the cytoplasm and translocates to the nucleus where it binds to either one of two transcription factor families, transcription factor (TCF) and lymphoid enhancer binding factor (LEF)1. This binding in turn displaces the transcription repressors Groucho and CtBP (Daniels and Weis 2005) and leads to the recruitment of co-activators such as cAMP-response element binding protein (CREB) binding protein/p300 (Takemaru and Moon 2000). Activation of the Wnt pathway results in the transcription of a number of target genes such as *axin2*, *tcf-1* and *CD44*, among many more (Mikami et al. 2001; Salahshor and Woodgett 2005; van de Wetering et al. 1991). The

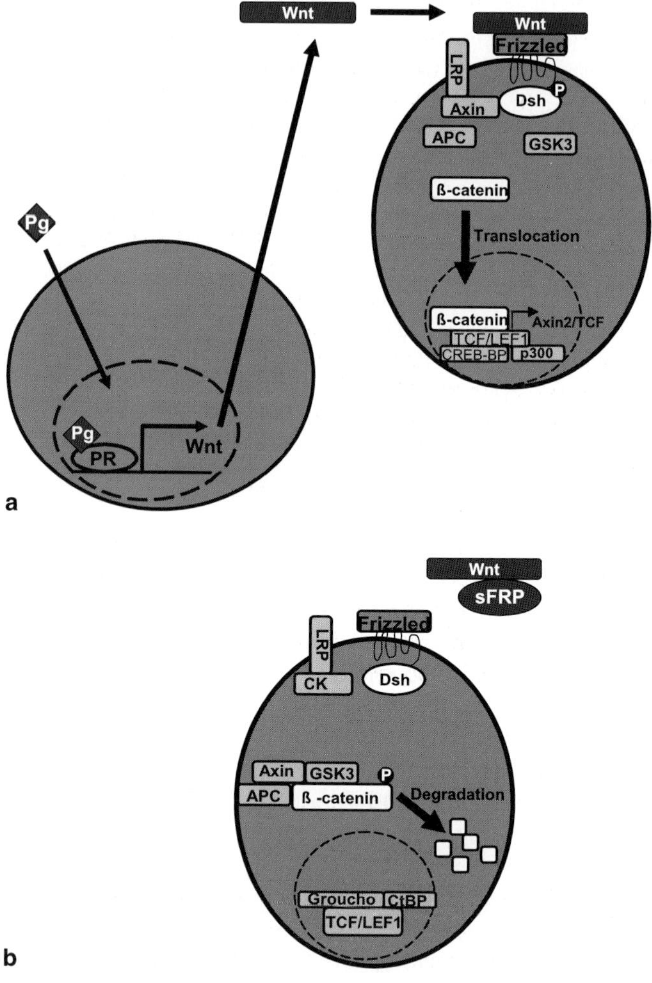

physiological response to Wnt signalling and activation of downstream targets is dependent upon the cellular context.

Competitive binding between Wnt ligands and secreted Frizzled-related protein or Wnt inhibitory factor is able to modulate the Wnt/β-

Fig. 2a, b. The Wnt/β-catenin signalling pathway. **a** In mammary epithelium, progesterone (Pg) binds to its receptor (PR) within the cell nucleus. Wnt proteins are in turn expressed, activated and secreted from the cell. These secreted extracellular Wnt proteins are then able to bind to the transmembrane receptor Frizzled and LRP5/6. This interaction results in the phosphorylation (P) of the cytoplasmic mediator Dishevelled, and the inhibition of GSK3β. When GSK3β is inactive, β-catenin accumulates within the cytoplasm followed by translocation to the nucleus where it binds to either TCF or LEF1. The transcription repressors Groucho and CtBP are displaced and co-activator CREB binding protein/p300 is recruited. Consequently, the transcription of downstream targets genes such as Axin2 and Lef1 is activated. **b** In the absence of Wnt ligands or the presence of Wnt inhibitors such as sFRP the pathway is no longer active. A complex of Axin/GSK3β/APC/β-catenin forms which allows the phosphorylation of β-catenin by GSK3β and casein kinase 1 (CK1). This targets β-catenin for degradation by the ubiquitin/proteasome pathway

Ocatenin pathway. The presence of inhibitory factors or absence of a Wnt activation results in the formation of an Axin/GSK3β/APC complex and phosphorylation of β-catenin at Ser residues 33, 37 and 45, and at Thr 41 by GSK3β and casein kinase 1 (CK1). This ultimately allows recognition by BTrCP and degradation by the ubiquitin/proteasome pathway.

The first mammalian Wnt gene, originally termed *Int-1*, was identified as a murine mammary tumour virus (MMTV) integration site in mammary tumours (Nusse and Varmus 1982). The Int-1 gene showed homology to the Drosophila segment polarity gene *Wingless* and subsequently *Int-1* and future family members were named Wnts.

To date, 16 mammalian Wnt genes have been identified and Wnt proteins can produce a wide variety of responses including cellular proliferation, differentiation, morphogenesis and cell fate decisions (Moon et al. 1997). The Wnt pathway has been implicated in a number of cancers with the classic example being colorectal cancer, where activating mutations within adenomatous polyposis coli (APC) result in dysregulation of β-catenin and the formation of intestinal polyps (Fodde et al. 2001). Other intestinal cancers have also been linked to mutations within *axin2* and the accumulation of β-catenin (Liu et al. 2000).

The Wnt pathway is also involved in both normal breast development and breast carcinomas. A number of studies have been conducted which demonstrate differential expression patterns of Wnt family members within mouse mammary development. Wnt-4, Wnt-5b, Wnt-6, Wnt-7b and Wnt-10b mRNA have been detected in mammary epithelium of various stages of mouse development. Wnt-4, Wnt-5b and Wnt-6 mRNA are upregulated during pregnancy and decrease with lactation. Wnt-10b expression can be detected from the early stages of embryonic mammary development and continues into puberty (Lane and Leder 1997). Wnt signalling plays a significant role in normal mammary gland development when expression begins at embryonic day 10.5 with the formation of two "mammary lines" (Veltmaat et al. 2003). In response to signals from the underlying mesenchyme, the mammary lines give rise to five mammary placodes that grow and invaginate the rudimentary fat pad. Wnt signalling coincides with mammary line development and localises in the mammary placodes and buds (Boras-Granic et al. 2006; Chu et al. 2004). Embryos transgenically engineered to over-express DKK1, an inhibitor of the Wnt pathway, display a complete absence of mammary placodes, demonstrating the importance of Wnt signalling in embryonic mammary development (Andl et al. 2002).

A link between mammary stem/progenitor cells and Wnt pathway activation-induced tumourigenesis has now been established. Transgenic expression of either Wnt1 or β-catenin results in widespread mammary hyperplasia and tumour formation (Imbert et al. 2001; Tsukamoto et al. 1988). The hyperplastic tissue contains increased numbers of stem/progenitor cells, which are thought to be directly responsible for transformed cells. Tumours that arise from stem/progenitor cells are heterogeneous showing cells of mixed lineage. Tumours that arise from Wnt activation also contain cells of both epithelial lineages (Li et al. 2003, 2004; Owens and Watt 2003; Shackleton et al. 2006). In MMTV–Wnt1 transgenic mice, loss of LRP5/6, a component of the Wnt pathway, results in a marked reduction in both the early proliferation of the progenitor cell population and formation of mammary tumours. Furthermore, LRP5$^{-/-}$ mammary cells were unable to reconstitute the full ductal trees through limiting dilution transplants (Lindvall et al. 2006).

Evidence is growing to support a role of the Wnt pathway in human mammary carcinomas since a number of Wnt pathway components are deregulated in human breast cancers. β-Catenin is stabilised, indicating activation of the Wnt pathway in 50% of breast carcinomas, which correlates with poor prognosis of the patient (Lin et al. 2000; Ryo et al. 2001). A number of Wnt ligands are upregulated in breast carcinomas and inhibitors of the pathway such as secreted Frizzled receptor protein (sFRP) are downregulated in breast carcinomas (Howe and Brown 2004; Ugolini et al. 1999, 2001). Ectopic expression of Wnt1 in human mammary epithelial cells elicits a DNA damage response which is an early event in human carcinogenesis (Bartkova et al. 2005; Gorgoulis et al. 2005), followed by Notch activation and tumourigenic transformation (Ayyanan et al. 2006). Most recently, treatment of human breast cancer cell lines with either Wnt ligands or inhibitors of the pathway showed that autocrine Wnt signalling contributes to breast cancer proliferation via activation of the canonical Wnt pathway, which utilises β-catenin, and also through EGFR transactivation (Schlange et al. 2007).

These data suggest that canonical Wnt signalling is essential for mammary SC activity. Therapies that specifically target the Wnt pathway in cancer SCs may be crucial to prevent recurrence and reduce breast cancer mortality rates.

10   Summary

The development of the breast is a complex process potentially involving multiple environmental and genetic factors. Hormonal steroids, including P and its downstream target Wnt4, play a key role in the development of both the normal breast and breast cancer. Signalling pathways such as Wnt are clearly implicated in both normal breast development and carcinomas through their regulation of the self-renewal and survival of mammary SCs. There are recent suggestions that CSCs, also known as cancer-initiating cells, may be the origin of breast tumours and responsible for breast cancer recurrence.

Strict control of the hormonal and SC signalling pathways is crucial for regulating SCs and for the correct development of the gland. Thus, deregulation of such pathways will strongly contribute to the formation of breast carcinomas. Consequently, the investigation of such pathways and the cross-talk between P and SC signalling pathways such as Wnt will be essential to the development of effective new anti-cancer drugs.

References

Al-Hajj M, Clarke MF (2004) Self-renewal and solid tumor stem cells. Oncogene 23:7274–7282

Al-Hajj M, Wicha MS, Benito-Hernandez A, Morrison SJ, Clarke MF (2003) Prospective identification of tumorigenic breast cancer cells. Proc Natl Acad Sci USA 100:3983–3988

Alvi AJ, Clayton H, Joshi C, Enver T, Ashworth A, Vivanco MM, Dale TC, Smalley MJ (2003) Functional and molecular characterisation of mammary side population cells. Breast Cancer Res 5:R1–8

Andl T, Reddy ST, Gaddapara T, Millar SE (2002) WNT signals are required for the initiation of hair follicle development. Dev Cell 2:643–653

Ayyanan A, Civenni G, Ciarloni L, Morel C, Mueller N, Lefort K, Mandinova A, Raffoul W, Fiche M, Dotto GP, Brisken C (2006) Increased Wnt signaling triggers oncogenic conversion of human breast epithelial cells by a Notch-dependent mechanism. Proc Natl Acad Sci USA 103:3799–3804

Bartkova J, Horejsi Z, Koed K, Kramer A, Tort F, Zieger K, Guldberg P, Sehested M, Nesland JM, Lukas C, Orntoft T, Lukas J, Bartek J (2005) DNA damage response as a candidate anti-cancer barrier in early human tumorigenesis. Nature 434:864–870

Bejsovec A (2005) Wnt pathway activation: new relations and locations. Cell 120:11–14

Bhanot P, Brink M, Samos CH, Hsieh JC, Wang Y, Macke JP, Andrew D, Nathans J, Nusse R (1996) A new member of the frizzled family from Drosophila functions as a Wingless receptor. Nature 382:225–230

Bonnet D (2003) Hematopoietic stem cells. Birth Defects Res C Embryo Today 69:219–229

Boras-Granic K, Chang H, Grosschedl R, Hamel PA (2006) Lef1 is required for the transition of Wnt signaling from mesenchymal to epithelial cells in the mouse embryonic mammary gland. Dev Biol 295:219–231

Boulanger CA, Wagner KU, Smith GH (2005) Parity-induced mouse mammary epithelial cells are pluripotent, self-renewing and sensitive to TGF-beta1 expression. Oncogene 24:552–560

Brisken C, Park S, Vass T, Lydon JP, O'Malley BW, Weinberg RA (1998) A paracrine role for the epithelial progesterone receptor in mammary gland development. Proc Natl Acad Sci USA 95:5076–5081

Brisken C, Heineman A, Chavarria T, Elenbaas B, Tan J, Dey SK, McMahon JA, McMahon AP, Weinberg RA (2000) Essential function of Wnt-4 in mammary gland development downstream of progesterone signaling. Genes Dev 14:650–654

Chang CF, Westbrook R, Ma J, Cao D (2007) Transforming growth factor-beta signaling in breast cancer. Front Biosci 12:4393–4401

Chen MS, Woodward WA, Behbod F, Peddibhotla S, Alfaro MP, Buchholz TA, Rosen JM (2007) Wnt/beta-catenin mediates radiation resistance of Sca1$^+$ progenitors in an immortalized mammary gland cell line. J Cell Sci 120:468–477

Chu EY, Hens J, Andl T, Kairo A, Yamaguchi TP, Brisken C, Glick A, Wysolmerski JJ, Millar SE (2004) Canonical WNT signaling promotes mammary placode development and is essential for initiation of mammary gland morphogenesis. Development 131:4819–4829

Clarke RB, Howell A, Potten CS, Anderson E (1997) Dissociation between steroid receptor expression and cell proliferation in the human breast. Cancer Res 57:4987–4991

Clarke RB, Anderson E, Howell A (2004) Steroid receptors in human breast cancer. Trends Endocrinol Metab 15:316–323

Clarke RB, Spence K, Anderson E, Howell A, Okano H, Potten CS (2005) A putative human breast stem cell population is enriched for steroid receptor-positive cells. Dev Biol 277:443–456

Clayton H, Titley I, Vivanco M (2004) Growth and differentiation of progenitor/stem cells derived from the human mammary gland. Exp Cell Res 297:444–460

Clemmons M, Gross P (2001) Estrogen and the risk of breast cancer. N Engl J Med 344:276–285

Daniels DL, Weis WI (2005) Beta-catenin directly displaces Groucho/TLE repressors from Tcf/Lef in Wnt-mediated transcription activation. Nat Struct Mol Biol 12:364–371

Deome KB, Faulkin LJ Jr, Bern HA, Blair PB (1959) Development of mammary tumors from hyperplastic alveolar nodules transplanted into gland-free mammary fat pads of female C3H mice. Cancer Res 19:515–520

Dexter TM, Spooncer E (1987) Growth and differentiation in the hemopoietic system. Annu Rev Cell Biol 3:423–441

Doble BW, Woodgett JR (2003) GSK-3: tricks of the trade for a multi-tasking kinase. J Cell Sci 116:1175–1186

Dontu G, Abdallah WM, Foley JM, Jackson KW, Clarke MF, Kawamura MJ, Wicha MS (2003) In vitro propagation and transcriptional profiling of human mammary stem/progenitor cells. Genes Dev 17:1253–1270

Dontu G, Jackson KW, McNicholas E, Kawamura MJ, Abdallah WM, Wicha MS (2004) Role of Notch signaling in cell-fate determination of human mammary stem/progenitor cells. Breast Cancer Res 6:R605–615

Dontu G, Wicha MS (2005) Survival of mammary stem cells in suspension culture: implications for stem cell biology and neoplasia. J Mammary Gland Biol Neoplasia 10:75–86

Ewan KB, Oketch-Rabah HA, Ravani SA, Shyamala G, Moses HL, Barcellos-Hoff MH (2005) Proliferation of estrogen receptor-alpha-positive mammary epithelial cells is restrained by transforming growth factor-beta1 in adult mice. Am J Pathol 167:409–417

Fillmore C, Kuperwasser C (2007) Human breast cancer stem cell markers CD44 and CD24: enriching for cells with functional properties in mice or in man? Breast Cancer Res 9:303

Fodde R, Smits R, Clevers H (2001) APC, signal transduction and genetic instability in colorectal cancer. Nat Rev Cancer 1:55–67

Fuchs E, Segre JA (2000) Stem cells: a new lease on life. Cell 100:143–155

Furth PA, Bar-Peled U, Li M (1997) Apoptosis and mammary gland involution: reviewing the process. Apoptosis 2:19–24

Gavin BJ, McMahon AP (1992) Differential regulation of the Wnt gene family during pregnancy and lactation suggests a role in postnatal development of the mammary gland. Mol Cell Biol 12:2418–2423

Goodell MA, Rosenzweig M, Kim H, Marks DF, DeMaria M, Paradis G, Grupp SA, Sieff CA, Mulligan RC, Johnson RP (1997) Dye efflux studies suggest that hematopoietic stem cells expressing low or undetectable levels of CD34 antigen exist in multiple species. Nat Med 3:1337–1345

Gorgoulis VG, Vassiliou LV, Karakaidos P, Zacharatos P, Kotsinas A, Liloglou T, Venere M, Ditullio RA Jr, Kastrinakis NG, Levy B, Kletsas D, Yoneta A, Herlyn M, Kittas C, Halazonetis TD (2005) Activation of the DNA damage checkpoint and genomic instability in human precancerous lesions. Nature 434:907–913

Hatsell S, Frost AR (2007) Hedgehog signaling in mammary gland development and breast cancer. J Mammary Gland Biol Neoplasia 12:163–173

Hiremath M, Lydon JP, Cowin P (2007) The pattern of {beta}-catenin responsiveness within the mammary gland is regulated by progesterone receptor. Development 134:3703–3712

Hovey RC, Trott JF, Vonderhaar BK (2002) Establishing a framework for the functional mammary gland: from endocrinology to morphology. J Mammary Gland Biol Neoplasia 7:17–38

Howe LR, Brown AM (2004) Wnt signaling and breast cancer. Cancer Biol Ther 3:36–41

Hu Y, Sun H, Drake J, Kittrell F, Abba MC, Deng L, Gaddis S, Sahin A, Baggerly K, Medina D, Aldaz CM (2004) From mice to humans: identification of commonly deregulated genes in mammary cancer via comparative SAGE studies. Cancer Res 64:7748–7755

Imbert A, Eelkema R, Jordan S, Feiner H, Cowin P (2001) Delta N89 beta-catenin induces precocious development, differentiation, and neoplasia in mammary gland. J Cell Biol 153:555–568

Johnston JB, Navaratnam S, Pitz MW, Maniate JM, Wiechec E, Baust H, Gingerich J, Skliris GP, Murphy LC, Los M (2006) Targeting the EGFR pathway for cancer therapy. Curr Med Chem 13:3483–3492

Jones PH (1997) Epithelial stem cells. Bioessays 19:683–690

Jordan CT, Lemischka IR (1990) Clonal and systemic analysis of long-term hematopoiesis in the mouse. Genes Dev 4:220–232

Kordon EC, Smith GH (1998) An entire functional mammary gland may comprise the progeny from a single cell. Development 125:1921–1930

Kritikou EA, Sharkey A, Abell K, Came PJ, Anderson E, Clarkson RW, Watson CJ (2003) A dual, non-redundant, role for LIF as a regulator of development and STAT3-mediated cell death in mammary gland. Development 130:3459–3468

Kuperwasser C, Chavarria T, Wu M, Magrane G, Gray JW, Carey L, Richardson A, Weinberg RA (2004) Reconstruction of functionally normal and malignant human breast tissues in mice. Proc Natl Acad Sci USA 101:4966–4971

Lane TF, Leder P (1997) Wnt-10b directs hypermorphic development and transformation in mammary glands of male and female mice. Oncogene 15:2133–2144

Laron Z, Pauli R, Pertzelan A (1989) Clinical evidence on the role of estrogens in the development of the breasts. Proc R Soc Edinburgh 95:13–22

Li Y, Welm B, Podsypanina K, Huang S, Chamorro M, Zhang X, Rowlands T, Egeblad M, Cowin P, Werb Z, Tan LK, Rosen JM, Varmus HE (2003) Evidence that transgenes encoding components of the Wnt signaling pathway preferentially induce mammary cancers from progenitor cells. Proc Natl Acad Sci USA 100:15853–15858

Liao MJ, Zhang CC, Zhou B, Zimonjic DB, Mani SA, Kaba M, Gifford A, Reinhardt F, Popescu NC, Guo W, Eaton EN, Lodish HF, Weinberg RA (2007) Enrichment of a population of mammary gland cells that form mammospheres and have in vivo repopulating activity. Cancer Res 67:8131–8138

Lin SY, Xia W, Wang JC, Kwong KY, Spohn B, Wen Y, Pestell RG, Hung MC (2000) Beta-catenin, a novel prognostic marker for breast cancer: its roles in cyclin D1 expression and cancer progression. Proc Natl Acad Sci USA 97:4262–4266

Lindvall C, Evans NC, Zylstra CR, Li Y, Alexander CM, Williams BO (2006) The Wnt signaling receptor Lrp5 is required for mammary ductal stem cell activity and Wnt1-induced tumorigenesis. J Biol Chem 281:35081–35087

Liu BY, McDermott SP, Khwaja SS, Alexander CM (2004) The transforming activity of Wnt effectors correlates with their ability to induce the accumulation of mammary progenitor cells. Proc Natl Acad Sci USA 101:4158–4163

Liu W, Dong X, Mai M, Seelan RS, Taniguchi K, Krishnadath KK, Halling KC, Cunningham JM, Boardman LA, Qian C, Christensen E, Schmidt SS, Roche PC, Smith DI, Thibodeau SN (2000) Mutations in AXIN2 cause colorectal cancer with defective mismatch repair by activating beta-catenin/TCF signalling. Nat Genet 26:146–147

Mallepell S, Krust A, Chambon P, Brisken C (2006) Paracrine signaling through the epithelial estrogen receptor alpha is required for proliferation and morphogenesis in the mammary gland. Proc Natl Acad Sci USA 103:2196–2201

Mikami T, Saegusa M, Mitomi H, Yanagisawa N, Ichinoe M, Okayasu I (2001) Significant correlations of E-cadherin, catenin, and CD44 variant form expression with carcinoma cell differentiation and prognosis of extrahepatic bile duct carcinomas. Am J Clin Pathol 116:369–376

Moon RT, Brown JD, Torres M (1997) WNTs modulate cell fate and behavior during vertebrate development. Trends Genet 13:157–162

Naccarato AG, Viacava P, Vignati S, Fanelli G, Bonadio AG, Montruccoli G, Bevilacqua G (2000) Bio-morphological events in the development of the human female mammary gland from fetal age to puberty. Virchows Arch 436:431–438

Nusse R, Varmus HE (1982) Many tumors induced by the mouse mammary tumor virus contain a provirus integrated in the same region of the host genome. Cell 31:99–109

Orkin SH (2000) Diversification of haematopoietic stem cells to specific lineages. Nat Rev Genet 1:57–64

Owens DM, Watt FM (2003) Contribution of stem cells and differentiated cells to epidermal tumours. Nat Rev Cancer 3:444–451

Parkin NT, Kitajewski J, Varmus HE (1993) Activity of Wnt-1 as a transmembrane protein. Genes Dev 7:2181–2193

Petersen OW, Hoyer PE, van Deurs B (1987) Frequency and distribution of estrogen receptor-positive cells in normal, nonlactating human breast tissue. Cancer Res 47:5748–5751

Phillips TM, McBride WH, Pajonk F (2006) The response of CD24$^{(-/low)}$/CD44^{+} breast cancer-initiating cells to radiation. J Natl Cancer Inst 98:1777–1785

Ponti D, Costa A, Zaffaroni N, Pratesi G, Petrangolini G, Coradini D, Pilotti S, Pierotti MA, Daidone MG (2005) Isolation and in vitro propagation of tumorigenic breast cancer cells with stem/progenitor cell properties. Cancer Res 65:5506–5511

Potten CS, Loeffler M (1990) Stem cells: attributes, cycles, spirals, pitfalls and uncertainties. Lessons for and from the crypt. Development 110:1001–1020

Proia DA, Kuperwasser C (2006) Reconstruction of human mammary tissues in a mouse model. Nat Protoc 1:206–214

Russo J, Ao X, Grill C, Russo IH (1999) Pattern of distribution of cells positive for estrogen receptor alpha and progesterone receptor in relation to proliferating cells in the mammary gland. Breast Cancer Res Treat 53:217–227

Ryo A, Nakamura M, Wulf G, Liou YC, Lu KP (2001) Pin1 regulates turnover and subcellular localization of beta-catenin by inhibiting its interaction with APC. Nat Cell Biol 3:793–801

Salahshor S, Woodgett JR (2005) The links between axin and carcinogenesis. J Clin Pathol 58:225–236

Schlange T, Matsuda Y, Lienhard S, Huber A, Hynes NE (2007) Autocrine WNT signaling contributes to breast cancer cell proliferation via the canonical WNT pathway and EGFR transactivation. Breast Cancer Res 9:R63

Schryver B, Hinck L, Papkoff J (1996) Properties of Wnt-1 protein that enable cell surface association. Oncogene 13:333–342

Seagroves TN, Krnacik S, Raught B, Gay J, Burgess-Beusse B, Darlington GJ, Rosen JM (1998) C/EBPbeta, but not C/EBPalpha, is essential for ductal morphogenesis, lobuloalveolar proliferation, and functional differentiation in the mouse mammary gland. Genes Dev 12:1917–1928

Shackleton M, Vaillant F, Simpson KJ, Stingl J, Smyth GK, Asselin-Labat ML, Wu L, Lindeman GJ, Visvader JE (2006) Generation of a functional mammary gland from a single stem cell. Nature 439:84–88

Sheridan C, Kishimoto H, Fuchs RK, Mehrotra S, Bhat-Nakshatri P, Turner CH, Goulet R Jr, Badve S, Nakshatri H (2006) CD44$^+$/CD24$^-$ breast cancer cells exhibit enhanced invasive properties: an early step necessary for metastasis. Breast Cancer Res 8:R59

Shipitsin M, Campbell LL, Argani P, Weremowicz S, Bloushtain-Qimron N, Yao J, Nikolskaya T, Serebryiskaya T, Beroukhim R, Hu M, Halushka MK, Sukumar S, Parker LM, Anderson KS, Harris LN, Garber JE, Richardson AL, Schnitt SJ, Nikolsky Y, Gelman RS, Polyak K (2007) Molecular definition of breast tumor heterogeneity. Cancer Cell 11:259–273

Siminovitch L, McCulloch EA, Till JE (1963) The distribution of colony-forming cells among spleen colonies. J Cell Physiol 62:327–336

Smalley MJ, Clarke RB (2005) The mammary gland "side population": a putative stem/progenitor cell marker? J Mammary Gland Biol Neoplasia 10:37–47

Stingl J, Eirew P, Ricketson I, Shackleton M, Vaillant F, Choi D, Li HI, Eaves CJ (2006) Purification and unique properties of mammary epithelial stem cells. Nature 439:993–997

Stylianou S, Clarke RB, Brennan K (2006) Aberrant activation of notch signaling in human breast cancer. Cancer Res 66:1517–1525

Takemaru KI, Moon RT (2000) The transcriptional coactivator CBP interacts with beta-catenin to activate gene expression. J Cell Biol 149:249–254

Travis RC, Key TJ (2003) Oestrogen exposure and breast cancer risk. Breast Cancer Res 5:239–247

Tsukamoto AS, Grosschedl R, Guzman RC, Parslow T, Varmus HE (1988) Expression of the int-1 gene in transgenic mice is associated with mammary gland hyperplasia and adenocarcinomas in male and female mice. Cell 55:619–625

Turashvili G, Bouchal J, Burkadze G, Kolar Z (2006) Wnt signaling pathway in mammary gland development and carcinogenesis. Pathobiology 73:213–223

Tworoger SS, Hankinson SE (2006) Prolactin and breast cancer risk. Cancer Lett 243:160–169

Ugolini F, Adelaide J, Charafe-Jauffret E, Nguyen C, Jacquemier J, Jordan B, Birnbaum D, Pebusque MJ (1999) Differential expression assay of chromosome arm 8p genes identifies Frizzled-related (FRP1/FRZB) and fibroblast growth factor receptor 1 (FGFR1) as candidate breast cancer genes. Oncogene 18:1903–1910

Ugolini F, Charafe-Jauffret E, Bardou VJ, Geneix J, Adelaide J, Labat-Moleur F, Penault-Llorca F, Longy M, Jacquemier J, Birnbaum D, Pebusque MJ (2001) WNT pathway and mammary carcinogenesis: loss of expression of candidate tumor suppressor gene SFRP1 in most invasive carcinomas except of the medullary type. Oncogene 20:5810–5817

van de Wetering M, Oosterwegel M, Dooijes D, Clevers H (1991) Identification and cloning of TCF-1, a T lymphocyte-specific transcription factor containing a sequence-specific HMG box. EMBO J 10:123–132

Veltmaat JM, Mailleux AA, Thiery JP, Bellusci S (2003) Mouse embryonic mammogenesis as a model for the molecular regulation of pattern formation. Differentiation 71:1–17

Villadsen R, Fridriksdottir AJ, Ronnov-Jessen L, Gudjonsson T, Rank F, LaBarge MA, Bissell MJ, Petersen OW (2007) Evidence for a stem cell hierarchy in the adult human breast. J Cell Biol 177:87–101

Watt FM (1998) Epidermal stem cells: markers, patterning and the control of stem cell fate. Philos Trans R Soc Lond B Biol Sci 353:831–837

Weber-Hall SJ, Phippard DJ, Niemeyer CC, Dale TC (1994) Developmental and hormonal regulation of Wnt gene expression in the mouse mammary gland. Differentiation 57:205–214

Wellings SR, Jensen HM, Marcum RG (1975) An atlas of subgross pathology of the human breast with special reference to possible precancerous lesions. J Natl Cancer Inst 55:231–273

Welm BE, Tepera SB, Venezia T, Graubert TA, Rosen JM, Goodell MA (2002) Sca-1(pos) cells in the mouse mammary gland represent an enriched progenitor cell population. Dev Biol 245:42–56

Woodward WA, Chen MS, Behbod F, Alfaro MP, Buchholz TA, Rosen JM (2007) WNT/beta-catenin mediates radiation resistance of mouse mammary progenitor cells. Proc Natl Acad Sci USA 104:618–623

Ernst Schering Foundation Symposium Proceedings, Vol. 1, pp. 25–43
DOI 10.1007/2789_2007_056
© Springer-Verlag Berlin Heidelberg
Published Online: 17 December 2007

Dynamic Regulation of Progesterone Receptor Activity in Female Reproductive Tissues

S.J. Han, F.J. DeMayo, B.W. O'Malley[✉]

Department of Molecular and Cellular Biology, Baylor College of Medicine, One Baylor Plaza, 77030 Houston, USA
email: *berto@bcm.tmc.edu*

Abstract. The progesterone receptor (PR) in cooperation with coregulator complexes coordinates crucial processes in female reproduction. To investigate the dynamic regulation of PR activity in vivo, a new transgenic mouse model utilizing a PR activity indicator (PRAI) system was generated. Studies utilizing the PRAI mouse have revealed that progesterone temporally regulates PR activity in female reproductive tissues. Specifically, progesterone rapidly enhances PR activity immediately after administration. However, chronic progesterone stimulation represses PR activity in female reproductive organs. Like progesterone, RU486 also temporally regulates PR activity in female reproductive organs. However, the temporal regulation of PR activity by RU486 is the inverse of progesterone's activity. RU486 acutely represses PR activity after injection but increases PR activity after chronic treatment in female reproductive tissues. Treatment with a mixed antagonist/agonist of PR, when compared to natural hormone, results in dramatically different tissue-specific patterns of intracellular PR activity, coregulator levels, and kinase activity. Transcriptional regulation of gene expression by PR is facilitated by coordinate interactions with the steroid receptor coactivators (SRCs). Bigenic PRAI–SRC knockout mouse models enabled us to draw a tissue-specific coactivator atlas for PR activity in vivo. Based on this atlas, we conclude that the endogenous physiological function of PR in distinct tissues is modulated by different SRCs. SRC-3 is the primary coactivator for PR in the breast and SRC-1 is the primary coactivator for PR in the uterus.

The progesterone receptor (PR) is an essential factor in female reproduction (Lydon et al. 1995). Mice in which the PR gene has been ablated by homologous recombination in embryonic stem cells [$PR^{-/-}$ mice] show defects in all aspects of female fertility (Lydon et al. 1996). The uterus is one of the major PR target organs and a major source of the infertility of $PR^{-/-}$ female mice. The uterus of $PR^{-/-}$ mice is unable to support embryo implantation, and uterine stromal cells fail to respond to progesterone after experimental induction of the decidual response. The $PR^{-/-}$ uterus also shows an increase in hyperplasia of the luminal and glandular epithelial cells in response to chronic estrogen and progesterone treatment when the mitogenic action of estrogen is unopposed by progesterone. Thus, PR plays critical roles in the regulation of uterine cell proliferation and uterine function.

During neonatal development, mammary ducts grow extensively until they occupy almost the entire mammary fat pad by 8 weeks of age. However, the number of mammary ductal branches was substantially reduced in age-matched $PR^{-/-}$ mice in comparison to wildtype (Lydon et al. 1995; Xu et al. 1998). In addition, $PR^{-/-}$ mice exhibited less dichotomous and lateral side-branching and defective alveolar development compared to wildtype (Lydon et al. 1995). Therefore, PR also has crucial roles in mammary gland development in mouse.

Although studies using $PR^{-/-}$ mice revealed the critical importance of PR in vivo, we have been unable to investigate the temporal and spatial distribution of PR activity in vivo. Therefore, we created an animal model in which we could investigate the dynamic changes in PR activity in response to various hormonal stimuli in vivo.

1   The PRAI Mouse Is a Novel Animal Model to Investigate PR Activity in vivo

Most previous studies investigating the molecular mechanisms regulating PR activity have utilized transient cell transfection assays as the major molecular biology technique. Albeit valuable, the limitation of the transient transfection assay is that it relies on overexpression of the individual components in cell culture. Consequently, cultured cells do not generally mimic the stoichiometry or the appropriate cellular milieu of intact animal tissues. In addition, paracrine communication between epithelial and stromal cells in response to external stimuli is an essential cellular process involved in the physiological function of PR in female reproductive tissues. But this critical paracrine communication does not exist in cell-based experiments. In order to overcome these problems, we generated a novel transgenic PR activity indicator (PRAI) mouse model. The PRAI mouse model employs a bacterial artificial chromosome (BAC) containing a modified PR gene and a responsive reporter gene. A BAC approach is advantageous since it utilizes large fragments of genomic DNA with an average of 100–300 kb in size that can be cloned in bacterial vectors and stably propagated in bacteria. The large size of BAC clones permits faithful direct tissue-specific expression of heterologous genes in vivo in BAC-transgenic mice if the appropriate regulatory sequences of the gene are present (Gong et al. 2003; Heintz 2000). More interestingly, BAC clones can be easily and precisely modified by bacterial recombination. Modifications of BAC clones include

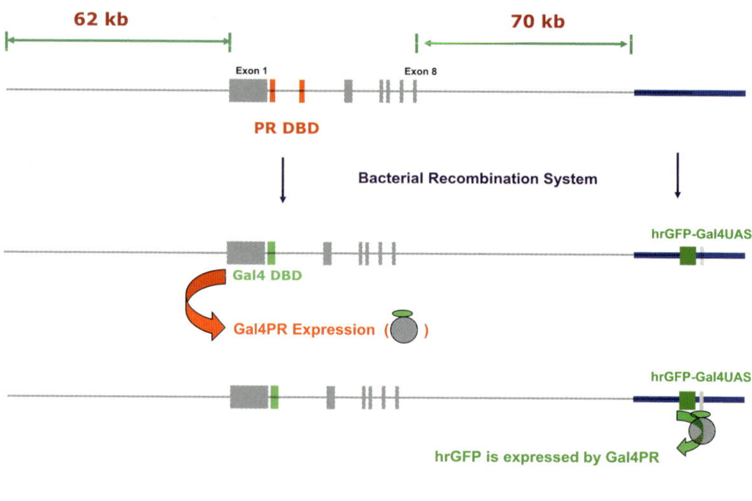

Fig. 1. PR BAC clone modification using bacterial recombination system. RPCI-23-422 I 15 PR BAC clone containing 62 kb of genomic DNA flanking 5′ to exon 1 and 70 kb of genomic DNA flanking 3′ to exon 8 of the PR gene is modified using bacterial recombination. The PR DNA binding domain (DBD) located from exon 2 to exon 3 in the PR gene was replaced with the Gal4 DBD generating the Gal4 version of PR (Gal4-PR). A second modification of this BAC clone included the incorporation of a reporter system consisting of five copies of UAS_G binding sites, a minimal promoter, and a hrGFP reporter gene. The Gal4-PR is recruited on the UAS_G binding sites to modulate hrGFP expression

single-base changes, deletions, and insertions (Ellis et al. 2001; Yu and Court 1998). In order to investigate PR activity in vivo, a modified PR BAC clone in which the DNA binding domain of the PR was replaced with the yeast Gal4 DNA binding domain (DBD) was generated using bacterial recombination. A humanized green fluorescent protein (hrGFP) reporter controlled by the upstream-activating sequences for the Gal4 gene (UAS_G) was inserted in tandem with the modified PR gene to monitor the activity of the Gal4 DBD-PR fusion (Gal4-PR) in vivo (Fig. 1). The PRAI mouse model was generated using this modified PR BAC clone.

In PRAI mice, the tissue-specific Gal4-PR expression pattern closely resembled endogenous PR expression in the major reproductive sites where the PR gene is expressed (Han et al. 2005). In addition to tissue-specific expression, the cell compartmental-specific expression of Gal4-PR in each tissue in response to different hormonal treatments also mimicked endogenous PR expression in female reproductive tissues (Han et al. 2005). Since the expression pattern of the Gal4-PR corresponded to the expression pattern of the endogenous PR, the PRAI mouse represents a unique animal model system to investigate PR activity in the female reproductive system in vivo.

2 Temporal Regulation of PR Activity in Female Reproductive Tissues in Response to Progesterone

Previous studies using cell transfection assays showed that progesterone enhanced PR-mediated gene transcription during progesterone treatment. However, the PRAI mouse system clearly demonstrated that progesterone does not always upregulate but instead dynamically modulates PR activity in progesterone target tissues such as the uterus and mammary gland (Table 1). For example, acute progesterone treatment (6 h following a single progesterone injection) significantly increased PR activity in the uterus of ovariectomized mice. In contrast, chronic progesterone treatment (3 days of daily progesterone injections) reduced PR activity in the uterus (Han et al. 2005). Microarray analysis for the murine uterine genes regulated by PR also presents temporal regulation of PR activity in uterus (Jeong et al. 2005). Why does chronic progesterone treatment reduce PR activity compared to acute progesterone treatment? First, PR gene expression is modulated by progesterone in female reproduction tissues (Table 2). For example, chronic progesterone treatment downregulates PR levels in both the uterus and mammary gland (Han et al. 2005, 2006). Therefore, reduced PR activity mediated by chronic progesterone treatment is associated with reduced PR expression. Second, in addition to regulating PR levels, progesterone can regulate levels of PR coregulators to modulate the ratio of coactivators to corepressors in a tissue-specific manner (Table 3). For example, chronic, but not acute, progesterone treatment increased levels of nuclear receptor co-repressor (NcoR) and silencing mediator for retinoid and thyroid hormone receptor (SMRT) in the stromal compartment of the uterus without changing coactivator levels. Therefore, chronic progesterone reduces the ratio of coactivators to corepressors in the uterus, preventing enhancement of PR activity (Han et al. 2005).

Collectively, progesterone differentially regulates PR activity, depending on the duration of treatment, by creating distinct cellular mi-

Table 1 Temporal and spatial changes of PR activity in uterus and mammary gland during progesterone or RU486 treatment

		Uterus			Mammary Gland
Hormone Treatment	LE	GE	S	M	LECs
Acute Progesterone	▲	▲	-	-	▲
Chronic Progesterone	▽	▽	▽	▽	▽
Acute RU486	▽	▽	-	-	▲
Chronic RU486	▲	▲	▲	▲	▲

GE, Glandular Epithelium; LE, Luminal Epithelium; LECs, Luminal Epithelial Cells; M, Myometrium; S, Stroma; ▲, increased PR activity compared to oil; ▽, decreased PR activity compared to oil; -, no changed PR activity compared to oil (Han et al. 2007)

Table 2 Temporal and spatial changes of PR expression in uterus and mammary gland during progesterone or RU486 treatment

		Uterus			Mammary Gland
Hormone Treatment	LE	GE	S	M	LECs
Acute Progesterone	-	-	-	-	-
Chronic Progesterone	▽	▽	▲	-	▽
Acute RU486	▽	▽	-	-	-
Chronic RU486	▲	-	▲	▲	▲

GE, Glandular Epithelium; LE, Luminal Epithelium; LECs, Luminal Epithelial Cells; M, Myometrium; S, Stroma; ▲, increased PR level compared to oil; ▽, decreased PR level compared to oil; -, no changed PR level compared to oil (Han et al. 2007)

Table 3 Temporal and spatial changes of PR coregulators expressions in uterus and mammary gland during progesterone or RU486 treatment

		Uterus			Mammary Gland
Hormone Treatment[a]	LE	GE	S	M	LECs
SRC-1 level					
Chronic Progesterone	▽	▽	▲	-	▲
Chronic RU486	▲	▲	▲	▲	-
SRC-2 level					
Chronic Progesterone	-	-	-	-	▲
Chronic RU486	-	-	-	-	▲
SRC-3 level					
Chronic Progesterone	▽	▽	▽	▽	▽
Chronic RU486	-	-	-	-	▲
NcoR level					
Acute Progesterone	▽	▽	-	-	-
Chronic Progesterone	▽	▽	▲	-	-
Acute RU486	▲	▲	▲	-	-
Chronic RU486	▽	▽	▽	-	-

[a]Expression levels of SRCs in both uterus and mammary gland are not changed during acute hormone treatment. GE, Glandular Epithelium; LE, Luminal Epithelium; LECs, Luminal Epithelial Cells; M, Myometrium; S, Stroma; ▲, increased PR level compared to oil; ▽, decreased PR level compared to oil; -, no changed PR level compared to oil (Han et al. 2007)

lieus of coregulators. Unfortunately, it is still unclear precisely how progesterone differentially modulates PR levels and the ratio of coactivators to corepressors in response to hormonal stimuli.

3   Dynamic Regulation of PR Function by RU486 in Reproductive Organs

In order to dynamically modulate PR activity, numerous synthetic PR ligands have been developed because of the physiological importance of PR in women's health. These synthetic PR ligands exhibit a spectrum of activity and range from more pure progesterone antagonists, such as onapristone (Michna et al. 1989) and ZK137316 (Teng et al. 2003), to mixed agonist/antagonists, which are currently known as selective progesterone receptor modulators (SPRMs), such as 11β-benzaldoxime-substituted estratrienes (Elger et al. 2000), some which already have been used for the clinical purposes of tissue-specific modulation (Chabbert-Buffet et al. 2005; Chwalisz et al. 2005). It is important that those regulatory molecules that modulate hormone responsiveness are investigated thoroughly to understand the potential mechanism and consequences of therapy with synthetic mixed agonists/antagonists.

RU486 (mifepristone), one type of SPRM, is a well-characterized antagonist of PR function. RU486 binds to PR and acutely impairs its gene regulatory activity (Baulieu 1991). For this reason, RU486 has been used clinically to prevent PR-dependent cellular processes. It is a well-known contraceptive and abortive agent (Baird et al. 2003; Baulieu 1997; Cheon et al. 2004). In addition to its PR-antagonistic activity, RU486 also has partial PR-agonist activity (Meyer et al. 1990; Wagner et al. 1996). For example, RU486 has been clinically used to treat uterine myoma and endometriosis because of its PR-agonistic antiproliferative and antiovulatory effects in a number of species including humans (Collins and Hodgen 1986; Spitz et al. 1996). In order to investigate the dynamic regulation of PR activity by RU486 in vivo, the effect of RU486 on PR activity was investigated using PRAI mice (Han et al. 2007).

4   Temporal Effects of RU486 on PR Activity

RU486 modulates PR activity in a complex manner, with outcomes depending on the specific tissue compartments and the duration of RU486 treatment (Table 1). For example, acute RU486 treatment results in PR-antagonistic activity in the uterus (Han et al. 2007). For this reason, RU486 has been extensively used as an abortion pill. Unlike the acute effects, chronic RU486 treatment enhances uterine PR activity (Han et al. 2007). PR-agonistic activity for RU486 also has been reported in clinical studies. For example, in post-menopausal women chronically treated with estradiol benzoate and mifepristone (RU486) at the dose of 100–200 mg/day, secretory transformation of the endometrium was observed because RU486 appears to function as an agonist (Gravanis et al. 1985).

In summary, RU486 has temporal PR regulatory activity depending on the duration of RU486 treatment. But the temporal regulation of PR activity by RU486 is inversely correlated with the effects of progesterone. These new findings regarding the in vivo effects of RU486 on PR activity should provide important information relative to hormone replacement therapy in postmenopausal women.

5 RU486 Activates Distinct Combinations of MAP Kinase Signaling Pathways in a Tissue-Specific Manner

The temporal effect of RU486 on PR activity raises the question as to how chronic RU486 treatment specifically increases rather than inhibits PR activity in female reproductive tissues. Steroid hormones may trigger cell type-specific kinase signaling pathways to modulate cellular processes. For example, estrogen activates multiple signaling pathways, including the mitogen-activated protein kinase (MAPK) pathway in endometrium to modulate apoptosis, proliferation, and inflammation (Seval et al. 2006). In addition to estrogen, RU486 also triggers specific kinase signaling to regulate PR-mediated cellular process. For example, RU486-induced labor has been associated with an increase in the active phosphorylated form of extracellular signal-regulated kinase ERK2 to increase contractility in vitro (Li et al. 2004). Interestingly, different combinations of MAPKs were activated by RU486, depending upon the tissue examined (Han et al. 2007). For example, ERK1/2 and c-Jun N-terminal kinase (JNK) pathways were activated by chronic RU486 treatment in the uterus. However, only the JNK signaling pathway was activated by chronic RU486 treatment in the luminal epithelial cells of the mammary gland. Therefore, RU486 triggers specific sets of MAPK pathways in each tissue to modulate PR activity in a tissue-specific manner. Like RU486, progesterone also can activate MAPK signaling pathways in both tissues. However, unlike RU486, chronic progesterone enhanced different MAPK pathways in female reproduction tissues. For example, the p38 kinase pathway was activated by chronic progesterone treatment, but not by chronic RU486 treatment in the mammary gland. Therefore, distinct combinations of activated MAPK signaling pathways induced by RU486 are different from those activated by progesterone in female reproductive organs. Using different sets of activated MAPKs, RU486 may alter distinct tissue-specific signaling programs that modify its agonist or antagonist actions in that tissue.

6   RU486 Modulates PR Activity Through MAP Kinase Signaling Pathways

PR is known to be phosphorylated by various kinases including CK-2, CDK2, and ERK1/2 in response to external hormonal stimuli (Knott et al. 2001; Lange et al. 2000; Zhang et al. 1994). Chronic RU486 treatment results in the activation of distinct combinations of MAPK pathways depending on tissue type. RU486-activated MAPKs can phosphorylate PR to modulate its activity. For example, in T47D cells, RU486 becomes a progesterone agonist in the presence of activators of protein kinase A (PKA), such as 8-bromo-cAMP, due to the loss of association of PR with corepressors, such as NCoR and SMRT (Beck et al. 1993; Fuhrmann et al. 2000; Sartorius et al. 1993). Therefore, it is possible that chronic RU486 and progesterone generate different, modified forms of PR as a result of differential effects on MAPK activity.

In addition to PR, coactivators are also phosphorylated by active MAPKs. For example, steroid receptor coactivator (SRC)-1 has been shown to be phosphorylated by ERK1/2 (Rowan et al. 2000b). Also, several different kinases such as JNK, glycogen synthase kinase (GSK3), p38, PKA, ERK, and IKKs can phosphorylate SRC-3 in response to different stimuli (Wu et al. 2004). Phosphorylation of SRCs can change their properties and allow for a more highly dynamic regulation of gene activity. For example, activated PKA signaling by 8-bromo-cAMP induces phosphorylation of two sites in SRC-1 to facilitate progesterone-independent activation of the chicken progesterone receptor (Rowan et al. 2000a). During retinoic acid-dependent activation of the retinoic acid receptor RARα in cultured cells, SRC-3 is phosphorylated by p38 MAPK, and this phosphorylation is required for optimal RARα-dependent gene regulation (Gianni et al. 2006). Therefore, chronic RU486 treatment triggers a specific combination of MAPK signaling pathways that modify PR coregulators in a manner that enhances their PR coactivator activity and thus has a positive effect on PR activity.

Ligand-dependent downregulation that leads to rapid and extensive loss of nuclear receptor protein is well recognized. In this process, different phosphorylation pathways may affect the proteasome-mediated degradation of nuclear receptors including estrogen receptor (ER)-α (Marsaud et al. 2003) and PR (Lange et al. 2000). By modulating MAPK activity, RU486 may influence the stability of PR and its coregulators because chronic RU486 treatment increases PR, SRC-1, and SRC-3 protein levels in female reproduction organs (Tables 2 and 3) (Han et al. 2007). Although not yet proved in a live animal model, it is possible that chronic RU486 treatment possesses agonist activity due to its ability to stabilize coactivator proteins as a result of its ability to influence endogenous kinases and thus coactivator phosphorylation. As a result, RU486 could alter PR activity in a tissue-specific manner through increased coactivator activity.

RU486 can increase the concentration of coactivator proteins in both a temporally and spatially distinct pattern in PR-target tissues. These findings confirm that the biological activity of mixed antagonist/agonist compounds is complex and not completely predictable. The activation of MAPK signaling pathways and upregulation of coactivators during chronic treatment with RU486 may broadly affect the biology of uterine and breast tissues. Thus, the PRAI mouse model can be used to generate an atlas of in vivo PR activity in response to progesterone and RU486. This atlas of endogenous PR activity should shed light on new ways to improve hormone replacement therapy for postmenopausal women.

7   A Tissue-Specific Partnership Exists Between PR and SRCs in Female Reproductive Organs

Studies using *PR$^{-/-}$* mice have shown that PR is essential for the proper development and function of all female reproductive organs, including the uterus and mammary gland (Lydon et al. 1995). Genetic deletion of the SRC family members in mice resulted in multiple abnormalities in the female reproductive system (Gehin et al. 2002; Mukherjee et al. 2006; Xu et al. 1998, 2000). The similarity of female reproductive defects between PR- and SRC-null mice implies that SRCs are involved in PR-mediated physiological process in the female reproductive system. However, it was still unclear whether SRC family members have cell- or tissue-specific roles in modulating PR activity in response to hormonal stimuli. To address this question, two novel bigenic mouse models, PRAI–SRC-1$^{-/-}$ and PRAI–SRC-3$^{-/-}$, were generated (Han et al. 2006). These mouse model systems were utilized to determine the contribution of each SRC to PR activity in different tissues. Thus, an atlas of PR activity was generated indicating the contributions of each SRC to PR activity in each reproductive tissue.

8   Uterine PR Requires SRC-1

Both PR and SRC-1 in the uterine stroma are important for the implantation process because ablation of either of these genes impairs the decidual response (Lydon et al. 1995; Xu et al. 1998). In addition, expression of both PR and SRC-1 in the stromal compartment of the uterus is highly elevated during decidualization (Han et al. 2006). However, to date there is no experimental evidence to support PR and SRC-1 functionally interacting to mediate the uterine decidual response. The PRAI–SRC-1$^{-/-}$ bigenic mouse model revealed that PR activity in the stromal compartment of the SRC-1$^{-/-}$ uterus was significantly impaired in response to chronic estrogen plus progesterone treatment compared to wildtype. Therefore, SRC-1 plays a significant role in the PR-dependent uterine stroma decidual response.

In addition to the uterus, defective mammary gland development is evident in both SRC-1$^{-/-}$ and $PR^{-/-}$ mice (Lydon et al. 1995; Xu et al. 1998). These similar defects in mammary gland development imply that SRC-1 may also be involved in PR-regulated mammary gland development. However, PR activity was not impaired in the LECs of the mammary gland of SRC-1$^{-/-}$ mice in response to chronic estrogen plus progesterone treatment. In addition to hrGFP reporter expression in PRAI–SRC-1$^{-/-}$ mice, the induction of other endogenous PR target genes, such as *Wnt-4* and *Amphiregulin*, was seen in the LECs of mammary tissue of

SRC-1$^{-/-}$ mice in response to chronic estrogen plus progesterone treatment. Therefore, despite the fact that SRC-1 colocalizes with PR in the LECs in response to hormonal treatment, the PR-dependent cellular response in mammary tissue is not impaired in SRC-1$^{-/-}$ mice. However, this observation raises the question as to why SRC-1$^{-/-}$ mice have defects in mammary gland development (Xu et al. 1998). In addition to PR, other nuclear receptors such as ER also have a crucial role in mammary gland development. For example, ER-$\alpha^{-/-}$ mice have defects in both ductal and alveolar development of mammary gland (Bocchinfuso and Korach 1997). Therefore, one likely possibility is that SRC-1 modulates a number of other steroid receptor signaling pathways, such as are involved in ER-α signaling, to regulate mammary gland development.

9 SRC-3 Is Involved in PR-Mediated Gene Regulation in the LECs of Mammary Gland

Unlike SRC-1, SRC-3 expression in the uterus is too low to be detected (Xu et al. 2000). In addition, the PR-mediated decidual response was not impaired in the uterus of SRC-3$^{-/-}$ mice and uterine stroma PR activity in bigenic PRAI–SRC-3$^{-/-}$ mice was unaltered following chronic estrogen plus progesterone treatment. Collectively, SRC-3 appears not to be a major contributor to PR-mediated uterine cellular functions in mice.

In contrast to the uterus, SRC-3 was expressed in the LECs during mammary gland development (Kuang et al. 2004; Xu et al. 2000). Inactivation of SRC-3 in mice partially attenuated mammary ductal growth during puberty and lobular alveolar development in response to estrogen and progesterone treatment (Kuang et al. 2004; Xu et al. 2000). In addition to normal mammary gland development, SRC-3 plays a critical role in mammary tumorigenesis mediated by pathways involving membrane-associated tyrosine kinase receptors and their downstream protein kinases (Kuang et al. 2005). PR activity was significantly reduced in the LECs of the mammary glands of PRAI–SRC-3$^{-/-}$ mice in response to chronic estrogen plus progesterone treatment. Expression of hrGFP and endogenous PR target genes (*Wnt-4* and *Amphiregulin*) in the LECs of the mammary gland are significantly reduced in the LECs of SRC-3$^{-/-}$ mice compared to wildtype. Therefore, the defect in mammary gland development in SRC-3$^{-/-}$ mice is likely due to a disconnection in the functional interaction between SRC-3 coactivation and PR-dependent cellular activity. In summary, SRC-3 is a key coactivator modulating PR-dependent gene expression in the mammary gland.

10   Factors Involved in Tissue-Specific Functional Interactions Between PR and Specific SRCs

According to the mouse GeneAtlas (Accelrys, San Diego) database (Mouse GeneAtlas GNF1M, MAS5 http://symatlas.gnf.org/SymAtlas/), each SRC family member has a tissue-specific expression pattern in mice. For example, SRC-1 mRNA is highly expressed in oocytes and a high level of SRC-3 mRNA is detected in mammary gland and placenta. This differential tissue-specific expression of coactivators is thought to contribute to the tissue specificity of in vivo coactivator functions for nuclear receptors. For example, SRC-1 levels in the uterine stroma were highly induced by chronic estrogen plus progesterone treatment in concert with increased PR expression. However, the SRC-3 level was too low to be detected in this uterine compartment under the same hormonal treatment. Therefore, SRC-1, but not SRC-3, is a major PR coactivator that modulates the PR-dependent decidual response.

In contrast to the uterus, PR colocalized with both SRC-1 and SRC-3 in the LECs of the mammary gland in response to chronic estrogen plus progesterone treatment. Nevertheless, only SRC-3 was involved in PR activation in the mammary gland. Thus, it appears that in different cellular contexts, PR utilizes different coregulators for its activity. How does PR select specific coactivators in each tissue for its function? Studies in our laboratory as well as others have indicated that external signals, such as hormones, trigger posttranslational modifications (such as acetylation, methylation, phosphorylation, ubiquitination, and sumoylation) of both PR and its coactivators (Chauchereau et al. 2003; Feng et al. 2006; Lange et al. 2000; Wu et al. 2004; Zhang et al. 1994). These different modifications of PR, SRC-1, and SRC-3 are likely to cause nuclear receptors and coactivators to interact differently in each tissue. The final consequence of these different modifications is differential activation of target genes and thus regulation of different biological functions.

Transgenic animal models, such as the one developed and used in this study, may further our understanding of the complex regulation of gene expression by selective receptor modulators and by the tissue-specific interaction of PR and its coregulators. Such studies of PR activity in vivo should aid in the development of new synthetic ligands for chemotherapy and hormone replacement therapy.

References

Baird DT, Brown A, Cheng L, Critchley HO, Lin S, Narvekar N, Williams AR (2003) Mifepristone: a novel estrogen-free daily contraceptive pill. Steroids 68:1099–1105

Baulieu EE (1991) On the mechanism of action of RU486. Ann N Y Acad Sci 626:545–560

Baulieu EE (1997) RU 486 (mifepristone). A short overview of its mechanisms of action and clinical uses at the end of 1996. Ann N Y Acad Sci 828:47–58

Beck CA, Weigel NL, Moyer ML, Nordeen SK, Edwards DP (1993) The progesterone antagonist RU486 acquires agonist activity upon stimulation of cAMP signaling pathways. Proc Natl Acad Sci U S A 90:4441–4445

Bocchinfuso WP, Korach KS (1997) Mammary gland development and tumorigenesis in estrogen receptor knockout mice. J Mammary Gland Biol Neoplasia 2:323–334

Chabbert-Buffet N, Meduri G, Bouchard P, Spitz IM (2005) Selective progesterone receptor modulators and progesterone antagonists: mechanisms of action and clinical applications. Hum Reprod Update 11:293–307

Chauchereau A, Amazit L, Quesne M, Guiochon-Mantel A, Milgrom E (2003) Sumoylation of the progesterone receptor and of the steroid receptor coactivator SRC-1. J Biol Chem 278:12335–12343

Cheon YP, DeMayo FJ, Bagchi MK, Bagchi IC (2004) Induction of cytotoxic T-lymphocyte antigen-2beta, a cysteine protease inhibitor in decidua: a potential regulator of embryo implantation. J Biol Chem 279:10357–10363

Chwalisz K, Perez MC, DeManno D, Winkel C, Schubert G, Elger W (2005) Selective progesterone receptor modulator development and use in the treatment of leiomyomata and endometriosis. Endocr Rev 26:423–438

Collins RL, Hodgen GD (1986) Blockade of the spontaneous midcycle gonadotropin surge in monkeys by RU 486: a progesterone antagonist or agonist? J Clin Endocrinol Metab 63:1270–1276

Elger W, Bartley J, Schneider B, Kaufmann G, Schubert G, Chwalisz K (2000) Endocrine pharmacological characterization of progesterone antagonists and progesterone receptor modulators with respect to PR-agonistic and antagonistic activity. Steroids 65:713–723

Ellis HM, Yu D, DiTizio T, Court DL (2001) High efficiency mutagenesis, repair, and engineering of chromosomal DNA using single-stranded oligonucleotides. Proc Natl Acad Sci U S A 98:6742–6746

Feng Q, Yi P, Wong J, O'Malley BW (2006) Signaling within a coactivator complex: methylation of SRC-3/AIB1 is a molecular switch for complex disassembly. Mol Cell Biol 26:7846–7857

Fuhrmann U, Hess-Stumpp H, Cleve A, Neef G, Schwede W, Hoffmann J, Fritzemeier KH, Chwalisz K (2000) Synthesis and biological activity of a novel, highly potent progesterone receptor antagonist. J Med Chem 43:5010–5016

Gehin M, Mark M, Dennefeld C, Dierich A, Gronemeyer H, Chambon P (2002) The function of TIF2/GRIP1 in mouse reproduction is distinct from those of SRC-1 and p/CIP. Mol Cell Biol 22:5923–5937

Gianni M, Parrella E, Raska I Jr, Gaillard E, Nigro EA, Gaudon C, Garattini E, Rochette-Egly C (2006) P38MAPK-dependent phosphorylation and degradation of SRC-3/AIB1 and RARalpha-mediated transcription. EMBO J 25:739–751

Gong S, Zheng C, Doughty ML, Losos K, Didkovsky N, Schambra UB, Nowak NJ, Joyner A, Leblanc G, Hatten ME, Heintz N (2003) A gene expression atlas of the central nervous system based on bacterial artificial chromosomes. Nature 425:917–925

Gravanis A, Schaison G, George M, de Brux J, Satyaswaroop PG, Baulieu EE, Robel P (1985) Endometrial and pituitary responses to the steroidal antiprogestin RU 486 in postmenopausal women. J Clin Endocrinol Metab 60:156–163

Han SJ, Jeong J, DeMayo FJ, Xu J, Tsai SY, Tsai MJ, O'Malley BW (2005) Dynamic cell type specificity of SRC-1 coactivator in modulating uterine progesterone receptor function in mice. Mol Cell Biol 25:8150–8165

Han SJ, DeMayo FJ, Xu J, Tsai SY, Tsai MJ, O'Malley BW (2006) Steroid receptor coactivator (SRC)-1 and SRC-3 differentially modulate tissue-specific activation functions of the progesterone receptor. Mol Endocrinol 20:45–55

Han SJ, Tsai SY, Tsai MJ, O'Malley BW (2007) Distinct temporal and spatial activities of RU486 on progesterone receptor function in reproductive organs of ovariectomized mice. Endocrinology 148:2471–2486

Heintz N (2000) Analysis of mammalian central nervous system gene expression and function using bacterial artificial chromosome-mediated transgenesis. Hum Mol Genet 9:937–943

Jeong JW, Lee KY, Kwak I, White LD, Hilsenbeck SG, Lydon JP, DeMayo FJ (2005) Identification of murine uterine genes regulated in a ligand-dependent manner by the progesterone receptor. Endocrinology 146:3490–3505

Knott KK, McGinley JN, Lubet RA, Steele VE, Thompson HJ (2001) Effect of the aromatase inhibitor vorozole on estrogen and progesterone receptor content of rat mammary carcinomas induced by 1-methyl-1-nitrosourea. Breast Cancer Res Treat 70:171–183

Kuang SQ, Liao L, Zhang H, Lee AV, O'Malley BW, Xu J (2004) AIB1/SRC-3 deficiency affects insulin-like growth factor I signaling pathway and suppresses v-Ha-ras-induced breast cancer initiation and progression in mice. Cancer Res 64:1875–1885

Kuang SQ, Liao L, Wang S, Medina D, O'Malley BW, Xu J (2005) Mice lacking the amplified in breast cancer 1/steroid receptor coactivator-3 are resistant to chemical carcinogen-induced mammary tumorigencsis. Cancer Res 65:7993–8002

Lange CA, Shen T, Horwitz KB (2000) Phosphorylation of human progesterone receptors at serine-294 by mitogen-activated protein kinase signals their degradation by the 26S proteasome. Proc Natl Acad Sci U S A 97:1032–1037

Li Y, Je HD, Malek S, Morgan KG (2004) Role of ERK1/2 in uterine contractility and preterm labor in rats. Am J Physiol Regul Integr Comp Physiol 287:R328–R335

Lydon JP, DeMayo FJ, Funk CR, Mani SK, Hughes AR, Montgomery CA Jr, Shyamala G, Conneely OM, O'Malley BW (1995) Mice lacking progesterone receptor exhibit pleiotropic reproductive abnormalities. Genes Dev 9:2266–2278

Lydon JP, DeMayo FJ, Conneely OM, O'Malley BW (1996) Reproductive phenotypes of the progesterone receptor null mutant mouse. J Steroid Biochem Mol Biol 56:67–77

Marsaud V, Gougelet A, Maillard S, Renoir JM (2003) Various phosphorylation pathways, depending on agonist and antagonist binding to endogenous estrogen receptor-alpha (ERalpha), differentially affect ER(alpha) extractability, proteasome-mediated stability, and transcriptional activity in human breast cancer cells. Mol Endocrinol 17:2013–2027

Meyer ME, Pornon A, Ji JW, Bocquel MT, Chambon P, Gronemeyer H (1990) Agonistic and antagonistic activities of RU486 on the functions of the human progesterone receptor. EMBO J 9:3923–3932

Michna H, Schneider MR, Nishino Y, el Etreby MF (1989) The antitumor mechanism of progesterone antagonists is a receptor mediated antiproliferative effect by induction of terminal cell death. J Steroid Biochem 34:447–453

Mukherjee A, Soyal SM, Fernandez-Valdivia R, Gehin M, Chambon P, DeMayo FJ, Lydon JP, O'Malley BW (2006) Steroid receptor coactivator 2 is critical for progesterone-dependent uterine function and mammary morphogenesis in the mouse. Mol Cell Biol 26:6571–6583

Rowan BG, Garrison N, Weigel NL, O'Malley BW (2000a) 8-Bromo-cyclic AMP induces phosphorylation of two sites in SRC-1 that facilitate ligand-independent activation of the chicken progesterone receptor and are critical for functional cooperation between SRC-1 and CREB binding protein. Mol Cell Biol 20:8720–8730

Rowan BG, Weigel NL, O'Malley BW (2000b) Phosphorylation of steroid receptor coactivator-1. Identification of the phosphorylation sites and phosphorylation through the mitogen-activated protein kinase pathway. J Biol Chem 275:4475–4483

Sartorius CA, Tung L, Takimoto GS, Horwitz KB (1993) Antagonist-occupied human progesterone receptors bound to DNA are functionally switched to transcriptional agonists by cAMP. J Biol Chem 268:9262–9266

Seval Y, Cakmak H, Kayisli UA, Arici A (2006) Estrogen-mediated regulation of p38 mitogen-activated protein kinase in human endometrium. J Clin Endocrinol Metab 91:2349–2357

Spitz IM, Croxatto HB, Robbins A (1996) Antiprogestins: mechanism of action and contraceptive potential. Annu Rev Pharmacol Toxicol 36:47–81

Teng J, Wang ZY, Bjorling DE (2003) Progesterone induces the proliferation of urothelial cells in an epidermal growth factor dependent manner. J Urol 170:2014–2018

Wagner BL, Pollio G, Leonhardt S, Wani MC, Lee DY, Imhof MO, Edwards DP, Cook CE, McDonnell DP (1996) 16 alpha-substituted analogs of the antiprogestin RU486 induce a unique conformation in the human progesterone receptor resulting in mixed agonist activity. Proc Natl Acad Sci U S A 93:8739–8744

Wu RC, Qin J, Yi P, Wong J, Tsai SY, Tsai MJ, O'Malley BW (2004) Selective phosphorylations of the SRC-3/AIB1 coactivator integrate genomic responses to multiple cellular signaling pathways. Mol Cell 15:937–949

Xu J, Qiu Y, DeMayo FJ, Tsai SY, Tsai MJ, O'Malley BW (1998) Partial hormone resistance in mice with disruption of the steroid receptor coactivator-1 (SRC-1) gene. Science 279:1922–1925

Xu J, Liao L, Ning G, Yoshida-Komiya H, Deng C, O'Malley BW (2000) The steroid receptor coactivator SRC-3 (p/CIP/RAC3/AIB1/ACTR/TRAM-1) is required for normal growth, puberty, female reproductive function, and mammary gland development. Proc Natl Acad Sci U S A 97:6379–6384

Yu D, Court DL (1998) A new system to place single copies of genes, sites and lacZ fusions on the Escherichia coli chromosome. Gene 223:77–81

Zhang Y, Beck CA, Poletti A, Edwards DP, Weigel NL (1994) Identification of phosphorylation sites unique to the B form of human progesterone receptor. In vitro phosphorylation by casein kinase II. J Biol Chem 269:31034–31040

Ernst Schering Foundation Symposium Proceedings, Vol. 1, pp. 45–54
DOI 10.1007/2789_2008_075
© Springer-Verlag Berlin Heidelberg
Published Online: 29 February 2008

Progesterone Signaling in Mammary Gland Development

O.M. Conneely[✉], B. Mulac-Jericevic, R. Arnett-Mansfield

Department of Molecular and Cellular Biology, Baylor College of Medicine, One Baylor Plaza, 77030 Houston, USA
email: *orlac@bcm.tmc.edu*

Abstract. The mammary gland undergoes extensive epithelial expansion and differentiation during pregnancy, leading ultimately to the development of functional milk-producing alveolar lobules. This phase of mammary gland remodeling is controlled primarily by the cooperative interplay between hormonal signals initiated by both progesterone and prolactin. Abrogation of mammary epithelial expression of receptors for either one of the hormones results in failure of alveologenesis and an absence of pregnancy-induced tertiary ductal side branches in the case of progesterone receptor-null (PRKO) mammary glands. By combining gene array approaches to identify PR- and prolactin (PRL)-dependent downstream signaling pathways and by using genetic mouse models to address the consequences of abrogation and/or misexpression of potential downstream genes, recent studies have begun to illuminate key signaling pathways that mediate the morphogenic effects of these hormones during pregnancy-induced mammary gland remodeling. Analysis of deregulated expression of PR-dependent gene transcripts in PRKO mammary glands has revealed that convergence between progesterone and prolactin signaling occurs in part through progesterone-dependent induction of mammary epithelial PRL receptors to prime the mammary epithelium to respond to PRL. Additional genes activated by PRs encode epithelial paracrine growth factor signals that regulate ductal and alveolar epithelial proliferation and survival, lineage-restricted transcription factors that control luminal and alveolar cell fate establishment and maintenance, and gap junction proteins that play a critical role in alveolar morphogenesis by establishment of epithelial cell polarity. Finally, two distinct isoforms of PRs (PR-A and PR-B) are coexpressed in the mammary gland and display extensively overlapping but partially distinct gene regulatory properties in relaying the progesterone signal.

1   Introduction

With the exception of the emergence of a primitive mammary epithelial rudiment that is established in the midgestational embryo, the bulk of mammary gland development occurs postnatally in two distinct growth phases that are initiated at the onset of puberty and pregnancy respectively. At puberty, the combined action of estrogen (E) and locally acting growth factors regulates proliferation of the terminal end buds located at the distal ends of the ductal epithelium to promote ductal elongation and dichotomous branching toward the limits of the mammary fat pad. While the transcription regulatory functions of E can be mediated by two distinct intracellular receptors, estrogen receptor (ER)α and ERβ, the ERα receptor is both necessary and sufficient for regulation of postpubertal mammary ductal development (Bocchinfuso et al. 2000; Mueller et al. 2002) while the ERβ receptor is dispensable (Forster et al. 2002). At pregnancy, exposure to progesterone (P) and prolactin (PRL) results in extensive epithelial proliferation, increased dichotomous side branching, and differentiation of milk-filled alveolar lobules that fill the interductal spaces by late pregnancy. At weaning, removal of the suckling stimulus to lactate results in involution of the lobular alveolar system through apoptosis and matrix-degrading proteinase-mediated remodeling to ultimately resemble the general architecture of the prepregnant mammary gland and complete the developmental cycle.

2 Progesterone Receptors in Mammary Gland Development

Progesterone signaling is mediated by interaction of the hormone with two receptor isoforms, progesterone receptor (PR)-A and PR-B, that are generated from a single gene by translation at two distinct ATG signals. The receptor proteins are identical, with the exception of an amino-terminal extension of approximately 160 amino acids that only PR-B contains. The PR isoforms that both overlapping and distinct cell context-dependent transcriptional regulatory functions when activated by progesterone and mediate partially overlapping but distinct tissue-selective reproductive functions of progesterone, including in the mammary gland. Null mutation of both PR isoforms in PR knockout (PRKO) mice has demonstrated that PRs are primarily required for pregnancy-associated tertiary ductal side-branching and lobuloalveolar differentiation of the mammary epithelium. The mammary glands of PRKO mice failed to develop the pregnancy associated side-branching of the ductal epithelium with attendant lobular alveolar differentiation despite normal postpubertal mammary gland morphogenesis of the virgin mice (Brisken et al. 1998; Lydon et al. 1995).

Throughout postpubertal mammary gland development, PRs are expressed exclusively in the epithelium (Ismail et al. 2002; Seagroves et al. 2000; Sivaraman et al. 2001). Consistent with these observations, tissue transplantation approaches using wildtype and PRKO mouse tissue to produce mammary gland recombinants that were devoid of PR in either the stromal or epithelial compartments has provided strong support for the functional involvement of epithelial rather than stromal PRs in mediating mammary gland morphogenic responses to P (Brisken et al. 1998). Development of the mammary gland from the juvenile to adult state is associated with a change in pattern of expression of epithelial PRs from a uniform to nonuniform pattern becoming localized to a scattered subset of epithelial cells that are segregated from proliferating cells throughout the adult ductal epithelium (Grimm et al. 2002; Ismail et al. 2002). Although progesterone receptors are transcriptional targets of ER in reproductive tissues and the mammary gland, absence of ERα does not result in loss of PR expression in mammary epithelium, and when supplemented with PRL and progesterone, ERα-null mammary glands respond by inducing lateral side-branching and lobular alveolar differentiation (Bocchinfuso et al. 2000).

The segregation of PR-positive mammary epithelial cells from proliferating epithelial cells is a conserved feature in normal rodent and human mammary tissue (Clarke et al. 1997; Ismail et al. 2002; Seagroves et al. 2000). Such an expression pattern led to the prediction that regulation of epithelial cell proliferation by progesterone would occur through a paracrine mechanism whereby PRs residing in nonproliferating cells induce expression of a signal that promotes proliferation of neighboring receptor-negative cells in a paracrine manner in the normal mammary gland. Consistent with this prediction, experiments that mixed PRKO and wildtype mammary epithelial cells demonstrated that while PRKO mammary epithelium cannot undergo side-branching, the branching and differentiation defects can be overcome when PRKO cells are placed in close contact with PR positive cells (Brisken et al. 1998). Thus, although lacking PR positive cells, the PRKO mammary epithelium still retains those PR-negative cells that are responsive to PR-mediated paracrine signaling.

3   PR-Dependent Molecular Signaling Pathways in the Mammary Gland

Pregnancy-associated mammary gland morphogenesis is achieved through cooperative interactions between both progesterone and PRL signaling via their respective receptors. Like PRKO mice, deletion of PRL receptors (PRLR) in PRLR$^{-/-}$ mice is sufficient to arrest lobular alveolar differentiation (Ormandy et al. 1997). Analysis of the effects of progesterone receptor deletion on PRL signaling reveals that PR-dependent upregulation of PRLR provides a point of convergence between these two signaling pathways that primes the mammary epithelium to respond to PRL signals. In the absence of PRs, PRKO mice have elevated levels of PRL hormone but significantly decreased levels of PRLR in the mammary gland.

The signaling mechanisms that mediate progesterone-dependent lateral branching from established ductal epithelium remain poorly understood. Previous studies identified the secreted glycoprotein Wnt-4 as a potential PR target that is coexpressed in PR positive-cells, is regulated by P, and is essential for regulating ductal branching via paracrine regulation of proliferation (Brisken et al. 2000). However, unlike PRKO mice, the morphogenic defects in Wnt-4-null mice are overcome in late pregnancy, suggesting that additional PR-dependent signaling pathways also play a key role in this response. A second potential paracrine mediator of progesterone action is amphiregulin, a key regulator of branching morphogenesis (Troyer and Lee 2001) whose induction in the mammary gland is inhibited in PRKO mice.

Progesterone signaling is also essential for expansion and differentiation of alveolar progenitor cells. In addition to priming of alveolar epithelium to respond to PRL by PR-dependent induction of PRLR gene expression, PRs also regulate the expression of paracrine signals that promote proliferative expansion and survival of PR-negative alveolar epithelial cells. One such signal is RANK-L (receptor activator of NF-κB-ligand) whose expression is regulated by P in PR-positive cells and whose action on neighboring proliferating cells drives the expression of cyclin D1, an essential mediator of alveolar proliferation and differentiation (Sicinski et al. 1995). Using gene array profiling of differentially expressed transcripts between estrogen and progesterone (E/P)-treated wildtype and PRKO mice to identify novel downstream mediators of progesterone action, we have also identified several lineage-restricted transcription factors whose expression in the mammary gland is downregulated in the absence of PRs (Fig. 1). These include GATA 3, a critical regulator of luminal and alveolar epithelial cell differentiation, and its transcriptional target FoxA1 (Asselin-Labat et al. 2007; Kouros-Mehr et al. 2006). In addition, transcription of the ets transcription factor Elf5, an essential regulator of alveologenesis, is dependent on PRs (Harris et al. 2006; Zhou et al. 2005). Finally, we identified PR-dependent tight junction proteins that likely contribute to cell–cell adhesion and maintenance of epithelial cell polarity during alveologenesis including claudin-3, claudin-4, claudin-7, and connexin-26. The latter of these, connexin-26, plays a critical role in alveolar epithelial cell survival during lobuloalveolar development (Bry et al. 2004). It is notable that several of these PR-dependent genes are also deregulated in mammary epithelium lacking PRLR. Whether they are direct tran-

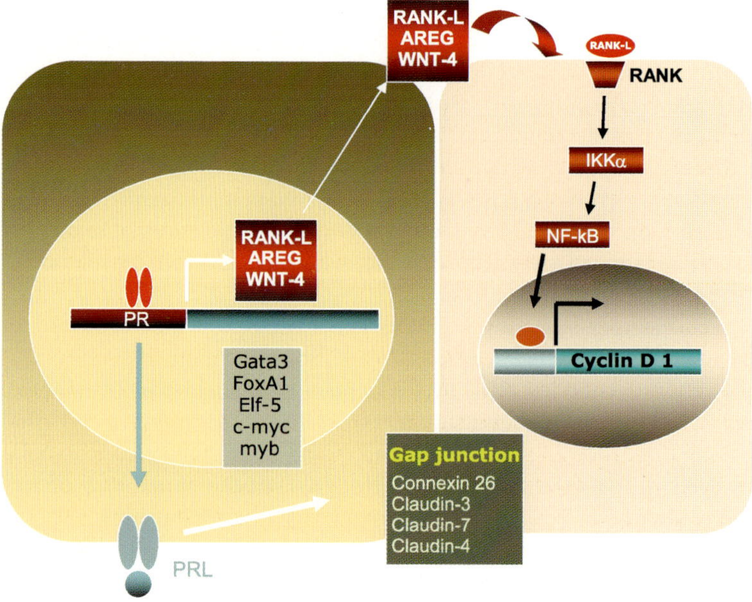

Fig. 1. Progesterone-dependent signaling pathways activated in the mammary gland during pregnancy

scriptional targets of both PR and PRLR or indirect downstream signals that are induced through PR-dependent upregulation of PRLR remains to be established.

4   PR Isoform-Selective Contribution to Pregnancy-Associated Mammary Gland Morphogenesis

Both isoforms of PR are coexpressed in mammary epithelial cells of the virgin mouse (Mote et al. 2006) and during pregnancy (Fantl et al. 1999), although the levels of PR-A protein exceed those of the PR-B isoform by at least a 2:1 ratio in both cases. To examine the selective contributions of each isoform to the morphogenic responses of the mammary epithelium to P, we analyzed the spatiotemporal expression patterns of the individual isoforms and compared the morphology of mammary glands of ovariectomized wildtype, PRAKO, and PRBKO mice after exposure to E and P. Ablation of PR-A in PRAKO mice did not affect the ability of PR-B to elicit normal P responsiveness in the mammary gland. The morphological changes in ductal side branching and lobular alveolar development in these glands were similar to those observed in wildtype mice (Mulac-Jericevic et al. 2000). Thus, PR-B is sufficient to elicit normal proliferation and differentiation of the mammary epithelium in response to P, and neither process appears to require functional expression of the PR-A protein. In contrast, analysis of the mammary glands of PRBKO mice has shown that, in the absence of PR-B, pregnancy-associated ductal side-branching and lobuloalveolar development in the mammary gland are markedly reduced because of decreased ductal and alveolar epithelial cell proliferation and decreased survival of alveolar epithelium (Mulac-Jericevic et al. 2003). Despite these defects, PR-A retains its normal segregated spatiotemporal pattern relative to proliferating cells in PRBKO mice and is expressed at a higher level than that observed for PR-B in PRAKO mice.

Interestingly, the PR isoform-selective morphogenic responses observed in PRAKO and PRBKO mice differ significantly from those observed when disruption of PR isoform ratios was achieved by overexpression of PR-A under the control of the cytomegalovirus promoter in transgenic mice (Shyamala et al. 1998). As a consequence of PR-A overexpression, mammary glands display increased ductal branching, hyperplasia, and disruption of basement membrane organization (Shyamala et al. 1998). Given the segregated expression pattern of PRs relative to proliferating cells in the normal gland, the striking differences in defects observed in PR-A transgenic relative to PRBKO$^{-/-}$ mice could be explained by inappropriate targeting in transgenic mice of PR isoform expression to epithelial subtypes that normally would not express these receptors, but may be competent to proliferate. Such targeting would breach the cellular segregation between receptor-expressing and proliferating cells observed in the normal gland, resulting in a scenario reminiscent of the inappropriate colocalization of steroid receptor expression and proliferation observed in mammary glands that have been exposed to carcinogen (Sivaraman et al. 2001) and in cells of breast tumors (Clarke et al. 1997). These findings suggest that disruption of the normal spatiotemporal expression pattern of PR-A may lead to aberrant regulation of proliferative target genes.

In an effort to elucidate the molecular genetic signaling pathways that are differentially regulated by individual PR isoforms in the mammary gland, we have examined the expression of a number of PR-dependent mammary epithelial gene transcripts (Fig. 1) in PRAKO and PRBKO mice. We have found that the individual isoforms regulate both overlapping and distinct progesterone-dependent genes in the mammary gland. While expression of amphiregulin, Wnt4, and several lineage-restricted transcription factors is unaffected by ablation of either isoform, the defects observed in PRBKO$^{-/-}$ mice are associated with a selectively reduced ability of the PR-A isoform to activate the RANK-L signaling pathway in response to P.

The decreased proliferative activity of PR-A in the mammary gland relative to PR-B may have important clinical implications with regard to the development of novel tissue-selective progestins for hormone therapy, as it suggests that PR-A-selective activation via PR-A-selective ligand agonists may limit adverse proliferative effects of progesterone in the mammary gland.

Acknowledgements. This work was supported by NIH grant No. HD32007 to O.M.C.

References

Asselin-Labat ML, Sutherland KD, Barker H, Thomas R, Shackleton M, Forrest NC, Hartley L, Robb L, Grosveld FG, van der Wees J, et al (2007) Gata-3 is an essential regulator of mammary-gland morphogenesis and luminal-cell differentiation. Nat Cell Biol 9:201–209

Bocchinfuso WP, Lindzey JK, Hewitt SC, Clark JA, Myers PH, Cooper R, Korach KS (2000) Induction of mammary gland development in estrogen receptor-alpha knockout mice. Endocrinology 141:2982–2994

Brisken C, Park S, Vass T, Lydon JP, O'Malley BW, Weinberg RA (1998) A paracrine role for the epithelial progesterone receptor in mammary gland development. Proc Natl Acad Sci USA 95:5076–5081

Brisken C, Heineman A, Chavarria T, Elenbaas B, Tan J, Dey SK, McMahon JA, McMahon AP, Weinberg RA (2000) Essential function of Wnt-4 in mammary gland development downstream of progesterone signaling. Genes Dev 14:650–654

Bry C, Maass K, Miyoshi K, Willecke K, Ott T, Robinson GW, Hennighausen L (2004) Loss of connexin 26 in mammary epithelium during early but not during late pregnancy results in unscheduled apoptosis and impaired development. Dev Biol 267:418–429

Clarke RB, Howell A, Potten CS, Anderson E (1997) Dissociation between steroid receptor expression and cell proliferation in the human breast. Cancer Res 57:4987–4991

Fantl V, Edwards PA, Steel JH, Vonderhaar BK, Dickson C (1999) Impaired mammary gland development in Cyl-1$^{(-/-)}$ mice during pregnancy and lactation is epithelial cell autonomous. Dev Biol 212:1–11

Forster C, Makela S, Warri A, Kietz S, Becker D, Hultenby K, Warner M, Gustafsson JA (2002) Involvement of estrogen receptor beta in terminal differentiation of mammary gland epithelium. Proc Natl Acad Sci USA 99: 15578–15583

Grimm SL, Seagroves TN, Kabotyanski EB, Hovey RC, Vonderhaar BK, Lydon JP, Miyoshi K, Hennighausen L, Ormandy CJ, Lee AV, et al (2002) Disruption of steroid and prolactin receptor patterning in the mammary gland correlates with a block in lobuloalveolar development. Mol Endocrinol 16:2675–2691

Harris J, Stanford PM, Sutherland K, Oakes SR, Naylor MJ, Robertson FG, Blazek KD, Kazlauskas M, Hilton HN, Wittlin S, et al (2006) Socs2 and elf5 mediate prolactin-induced mammary gland development. Mol Endocrinol 20:1177–1187

Ismail PM, Li J, DeMayo FJ, O'Malley BW, Lydon JP (2002) A novel LacZ reporter mouse reveals complex regulation of the progesterone receptor promoter during mammary gland development. Mol Endocrinol 16:2475–2489

Kouros-Mehr H, Slorach EM, Sternlicht MD, Werb Z (2006) GATA-3 maintains the differentiation of the luminal cell fate in the mammary gland. Cell 127:1041–1055

Lydon JP, DeMayo FJ, Funk CR, Mani SK, Hughes AR, Montgomery CA Jr, Shyamala G, Conneely OM, O'Malley BW (1995) Mice lacking progesterone receptor exhibit pleiotropic reproductive abnormalities. Genes Dev 9:2266–2278

Mote PA, Arnett-Mansfield RL, Gava N, deFazio A, Mulac-Jericevic B, Conneely OM, Clarke CL (2006) Overlapping and distinct expression of progesterone receptors A and B in mouse uterus and mammary gland during the estrous cycle. Endocrinology 147:5503–5512

Mueller SO, Clark JA, Myers PH, Korach KS (2002) Mammary gland development in adult mice requires epithelial and stromal estrogen receptor alpha. Endocrinology 143:2357–2365

Mulac-Jericevic B, Mullinax RA, DeMayo FJ, Lydon JP, Conneely OM (2000) Subgroup of reproductive functions of progesterone mediated by progesterone receptor-B isoform. Science 289:1751–1754

Mulac-Jericevic B, Lydon JP, DeMayo FJ, Conneely OM (2003) Defective mammary gland morphogenesis in mice lacking the progesterone receptor B isoform. Proc Natl Acad Sci USA 100:9744–9749

Ormandy CJ, Binart N, Kelly PA (1997) Mammary gland development in prolactin receptor knockout mice. J Mammary Gland Biol Neoplasia 2:355–364

Seagroves TN, Lydon JP, Hovey RC, Vonderhaar BK, Rosen JM (2000) C/EBPbeta (CCAAT/enhancer binding protein) controls cell fate determination during mammary gland development. Mol Endocrinol 14:359–368

Shyamala G, Yang X, Silberstein G, Barcellos-Hoff MH, Dale E (1998) Transgenic mice carrying an imbalance in the native ratio of A to B forms of progesterone receptor exhibit developmental abnormalities in mammary glands. Proc Natl Acad Sci USA 95:696–701

Sicinski P, Donaher JL, Parker SB, Li T, Fazeli A, Gardner H, Haslam SZ, Bronson RT, Elledge SJ, Weinberg RA (1995) Cyclin D1 provides a link between development and oncogenesis in the retina and breast. Cell 82:621–630

Sivaraman L, Hilsenbeck SG, Zhong L, Gay J, Conneely OM, Medina D, O'Malley BW (2001) Early exposure of the rat mammary gland to estrogen and progesterone blocks co-localization of estrogen receptor expression and proliferation. J Endocrinol 171:75–83

Troyer KL, Lee DC (2001) Regulation of mouse mammary gland development and tumorigenesis by the ERBB signaling network. J Mammary Gland Biol Neoplasia 6:7–21

Zhou J, Chehab R, Tkalcevic J, Naylor MJ, Harris J, Wilson TJ, Tsao S, Tellis I, Zavarsek S, Xu D, et al (2005) Elf5 is essential for early embryogenesis and mammary gland development during pregnancy and lactation. EMBO J 24:635–644

Ernst Schering Foundation Symposium Proceedings, Vol. 1, pp. 55–76
DOI 10.1007/2789_2007_057
© Springer-Verlag Berlin Heidelberg
Published Online: 17 December 2007

Steroid Receptor Coactivator 2: An Essential Coregulator of Progestin-Induced Uterine and Mammary Morphogenesis

A. Mukherjee, P. Amato, D. Craig-Allred, F.J. DeMayo,
B.W. O'Malley, J.P. Lydon(✉)

Department of Molecular and Cellular Biology, Baylor College of Medicine, One Baylor
Plaza, 77030 Houston, USA
email: *jlydon@bcm.tmc.edu*

Abstract. The importance of the progesterone receptor (PR) in transducing
the progestin signal is firmly established in female reproductive and mammary
gland biology; however, the coregulators preferentially recruited by PR in these
systems have yet to be comprehensively investigated. Using an innovative ge-

netic approach, which ablates gene function specifically in murine cell-lineages that express PR, steroid receptor coactivator 2 (SRC-2, also known as TIF-2 or GRIP-1) was shown to exert potent coregulator properties in progestin-dependent responses in the uterus and mammary gland. Uterine cells positive for PR (but devoid of SRC-2) led to an early block in embryo implantation, a phenotype not shared by knockouts for SRC-1 or SRC-3. In the case of the mammary gland, progestin-dependent branching morphogenesis and alveolo-genesis failed to occur in the absence of SRC-2, thereby establishing a critical coactivator role for SRC-2 in cellular proliferative programs initiated by pro-gestins in this tissue. Importantly, the recent detection of SRC-2 in both human endometrium and breast suggests that this coregulator may provide a new clini-cal target for the future management of female reproductive health and/or breast cancer.

1   Introduction

The progesterone receptor knockout (PRKO) mouse established the pro-gesterone receptor (PR) as an indispensable regulator of female repro-ductive function (Lydon et al. 1995). Abrogation of PR compromised murine uterine morphogenesis and directly blocked the normal opera-tion of the hypothalamo-pituitary–ovarian axis (reviewed in Fernandez-Valdivia et al. 2005). The PRKO also established a proliferative role for progestins in the mammary epithelium (Lydon et al. 1995), the pro-liferation of which leads to ductal side-branching and alveologenesis in the parous animal. Importantly, the PRKO mouse highlighted the essen-tial involvement of PR in mammary tumorigenesis (Lydon et al. 1999; Chatterton et al. 2002; Medina et al. 2003), a finding that supports con-clusions drawn from human observational studies (Colditz et al. 1993; Ross et al. 2000; Schairer et al. 2000), much publicized clinical trials (Rossouw et al. 2002; Beral 2003), and recent correlations made be-tween the decline (since 2002) in hormone therapy use in some patient populations in the United States and the contemporaneous reduction in newly diagnosed breast cancer cases (Clarke et al. 2006).

Despite major advances in the progestin field, two key questions have emerged from over a decade of PRKO studies: (1) what are the downstream molecular pathways (and networks) that transduce an external progestin signal to an appropriate physiological response in a given target cell, and (2) which coregulators (coactivators and/or corepressors) are selectively enlisted by PR to regulate the expression of these effector pathways in vivo? Addressing these interconnected questions promises to provide much needed insight into the molecular mechanisms that underpin tissue-selective responses to progestins in normal physiology as well as aid in the formulation of new conceptual frameworks by which to further investigate the physiological involvement of progestin-action in such clinical disorders as female infertility and breast cancer.

While progress has been attained in disclosing the transcriptome controlled by PR in select murine target tissues [for example, the uterus (Das et al. 1995; Lim et al. 1999; Brisken et al. 2000; Cheon et al. 2002; Takamoto et al. 2002; Ismail et al. 2004; Jeong et al. 2005)], identifying coregulators preferentially engaged in PR-mediated physiological effects is only now being realized due in large part to the recent development and deployment of innovative murine engineering methodologies.

This review describes how the application of such technology recently uncovered a pivotal role for the multifunctional coregulator, steroid receptor coactivator 2 (SRC-2), in progestin-initiated physiological processes in the murine uterus and mammary gland.

2   SRC-2 Is a Member of the Steroid Receptor Coactivator/p160 Family

Seminal in vitro investigations by the O'Malley group revealed that the transactivational potency of agonist-bound PR is enhanced by increased cellular levels of members of the SRC/p160 family (reviewed in McKenna and O'Malley 2002). In addition to SRC-2 (TIF-2/GRIP-1/NcoA-2), the SRC/p160 family comprises two additional members: SRC-1 (ERAP140/ERAP160/NcoA-1) and SRC-3 (p/CIP/RAC3/AIB1/ TRAM-1/ACTR/NcoA-3) (reviewed in Lonard and O'Malley 2005). To potentiate nuclear receptor (NR) transcriptional activation in response to agonist, each coactivator has been shown to directly contact—via specific LXXLL motifs within their centrally located NR interaction domain (Fig. 1)—the highly conserved activation 2 domain of NRs. Importantly, two activation domains (AD1 and AD2) located at the C-terminal region of all three coactivators are responsible for enlisting secondary coactivators (or co-coactivators). For example, AD1 has been shown to recruit the related histone acetyltransferases (HATs) p300 and the cyclic AMP-response element binding protein (CREB)-binding protein (CBP) whereas AD2 is known to enlist arginine methyltransferases such as coactivator-associated arginine methyltransferases I (CARM1) (reviewed in Lonard and O'Malley 2006). The histone-modifying activities of these secondary coactivators (coupled with weak intrinsic HAT activity of some SRC members) contribute to local chromatin remodeling that enables the general transcriptional machinery access to the promoter region of NR target genes. In addition to histones, these secondary coactivators are known to post-translationally modify other target proteins within the transcriptional complex such as other coactivators and transcription factors. The basic helix loop helix-Per/ARNT/Sim (bHLH-PAS) domain, the most conserved region among SRC members,

---

Fig. 1A,B. The SRC/p160 family of multifunctional coregulators. **A** The structural organization of human (*h*) SRC-1, -2, and -3 proteins. The basic loop helix, Per/ARNT/Sim, receptor interaction, and activation domains are denoted by *bHLH*, *PAS*, *RID*, and *AD* respectively. The amino acid region responsible for histone acetyl transferase (HAT) activity in SRC-1 and -3 is also indicated; *Q* denotes the glutamine-rich region. *Below*, the similarity and identity (*in parentheses*) of amino acid sequences within key functional domains of SRC members is displayed. Overall amino acid similarity and identity between SRC members is: hSRC1/2, 54% (46% identical); hSRC1/3, 50% (43%); and hSRC2/3, 55% (48%). Amino acid sequence alignments were conducted using *LALNVIEW* software (http://pbil.univ-lyon1.fr/software/lalnview.html; Duret et al. 1996). **B** An idealized schematic model in which the partnership of SRC-2 with PR at the genome constitutes part of a dynamic multiprotein transcriptional complex in which such secondary coregulators as p300 (Chen et al. 2000a), CARM-1 (Chen et al. 1999), FLASH (Kino et al. 2004), GAC63 (Chen et al. 2005), and CoCoA (Kim and Stallcup 2004). The complexes differentially assemble and disassemble depending on a particular input signal, such as a distinct phosphorylation event mediated by a growth factor or cell survival signal-induced kinase

located at the N-terminal region (Fig. 1), has also been shown to recruit secondary coregulators and transcription factors. In the case of SRC-2, these secondary coregulators include coiled-coil co-activator (CoCoA) (Kim et al. 2003), flightless-I (Fli-I) (Lee and Stallcup 2006), GRIP1-associated co-activator 63 (GAC63) (Chen et al. 2005), and the transcription factors myocyte-enhancer factor 2C (MEF-2C) (Chen et al. 2000b) and TEF4 (Belandia and Parker 2000). In other regulatory proteins, the bHLH-PAS motif has been implicated in both DNA and ligand binding (Huang et al. 1993; Gu et al. 2000), suggesting that this domain may subserve functions in SRCs beyond those currently recognized.

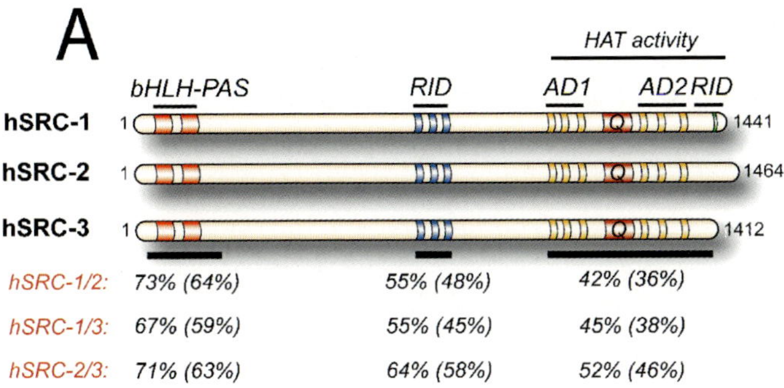

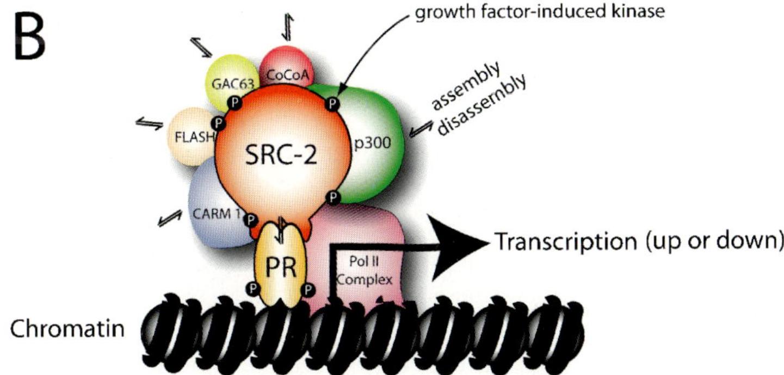

It is important to note that superimposed on the multitude of protein–protein interactions that allow SRC-2 to convey (and modulate) the signaling input dispatched from ligand-bound NR to the general transcriptional machinery, a multiplicity of interacting signaling inputs [i.e., phosphorylation events from extracellular growth and cell survival factors (Duong et al. 2006; Frigo et al. 2006)] are also being received and deciphered by SRC-2 within the multicomponent transcriptional complex. Moreover, while SRC-2 has been primarily considered a coactivator, recent studies provide compelling evidence that this protein can (within certain cellular contexts) exert repressor functions (Rogatsky et al. 2002), once again underscoring the versatility of this multifunctional coregulator.

Although in vitro studies were responsible for revealing the existence of the SRC family, experimental mouse genetics would disclose important physiologic roles for all three coactivators in progestin-initiated signaling events. In keeping with their multifunctional capabilities, mouse experiments have also uncovered essential roles for each SRC in important physiological processes outside the realm of progestin control.

3   Steroid Receptor Coactivator Members Are Complicit in Progestin-Dependent Physiological Processes—Insights from the Mouse

The finding that all SRC family members can directly interface with PR and that distinct SRC combinatorial assemblies can activate specific gene sets provided strong support for the argument that differential enlistment of SRCs constitutes an important mechanism by which the PR differentially mediates progestin effects in different target tissues. Although fertile and viable, the SRC-1 knockout (KO) mouse exhibits a limited decidual response in the uterus (Xu et al. 1998), suggesting that this coactivator (with others) is required for full elaboration of this morphogenetic response which depends on progestin exposure. The SRC-3 female is also fertile and viable but displays a partial block in hormone-induced mammary ductal side-branching and alveologenesis (Xu et al. 2000), epithelial changes that are absent in the PRKO gland (Lydon et al. 1995). Together, these studies support the assertion that

SRC-1 and -3 are recruited for a subset of PR-mediated transcriptional responses in the uterus and mammary gland respectively; recent studies with the PR activity indicator (PRAI) model provides further support for this proposal (Han et al. 2005; Han et al. 2006). Mouse investigations have also underscored important roles for SRC-1 and SRC-3 in areas of normal physiology and disease which are not directly controlled by progestins, these include: somatic cell growth (Wang et al. 2000; Xu et al. 2000), energy homeostasis (Louet et al. 2006; Wang et al. 2006), thyroid hormone function (Ying et al. 2005), bone-turnover (Modder et al. 2004), prostate (Zhou et al. 2005) and B-cell lymphoma (Coste et al. 2006) to name but a few.

Unlike KOs for SRC-1 and -3, the SRC-2 KO [termed: transcriptional intermediary factor 2 KO (TIF2$^{-/-}$)] exhibits significant reproductive defects in both sexes (Gehin et al. 2002). In the case of the female, absence of SRC-2 results in placental hypoplasia that underlies a severe hypofertility phenotype. Further investigations revealed that TIF2$^{-/-}$ pups (both sexes) are significantly underrepresented in litters from TIF2$^{+/-}$ crosses (A. Mukherjee, F.J. demayo, and J.P. Lydon, unpublished observations); TIF2$^{-/-}$ females resulting from such crosses are infertile. As for KOs for SRC-1 and -3, global removal of SRC-2 function results in additional physiological defects not directly related to male or female reproduction [i.e., a decrease in early postnatal survival (Mark et al. 2004), dysregulation in energy homeostasis (Jeong et al. 2006), and insidious adrenocortical insufficiency (Patchev et al. 2007)].

4   A Cell Lineage-Specific Ablation Approach Uncovers an Essential Coactivator Role for SRC-2 in Progestin-Dependent Tissue Responses in the Mouse

The placental defect exhibited by the global KO for SRC-2, in conjunction with the recent finding that a subset of murine cell-lineages that express PR also express SRC-2 (Mukherjee et al. 2006b), indicated that SRC-2 (like SRC-1 and -3) may play a crucial coactivator role in a subset of physiological responses that require PR activity. To test this proposal, a PR$^{Cre/+}$ SRC-2$^{flox/flox}$ bigenic mouse was generated (Mukherjee

Fig. 2A–D. Absence of uterine SRC-2 results in a block in embryo implantation and a partial decidual response. **A** *Arrows* indicate the position of implantation sites in the uterus (*1*) of a SRC-2$^{flox/flox}$ [or wild-type (WT)] mouse [5.5 days postcoitum (d.p.c.)]. However, implantation sites were not detected in uteri from similarly treated PR$^{Cre/+}$SRC-2$^{flox/flox}$ mice (*2*). The average number of implantation sites per genotype per total number of mice examined is tabulated. **B** The morphological response of the left (*L*) uterine horn to a deciduogenic stimulus for SRC-2$^{flox/flox}$ (*1*), PR$^{Cre/+}$SRC-2$^{flox/flox}$ (*2*), and PR$^{Cre/+}$SRC-2$^{flox/flox}$ SRC-1KO trigenic (*3*) mice is shown. The right (*R*) uterine horn represents the unstimulated control. Although the PR$^{Cre/+}$SRC-2$^{flox/flox}$ uterus (*2*) displays a limited decidual response, note the absence of a decidual response in the PR$^{Cre/+}$SRC-2$^{flox/flox}$ SRC-1 KO trigenic uterus (*3*). **C** Graphic representation of the average weight ratios (±SD) of stimulated (*L*) to control (*R*) horn for SRC-2$^{flox/flox}$ (*1*), PR$^{Cre/+}$SRC-2$^{flox/flox}$ (*2*), and PR$^{Cre/+}$SRC-2$^{flox/flox}$ SRC-1KO trigenic (*3*) uteri. **D** Western analysis reveals uterine tissue from untreated adult virgin SRC-2$^{flox/flox}$ (*1*) and PR$^{Cre/+}$SRC-2$^{flox/flox}$ (*2*) mice; they show equivalent levels of uterine SRC-1 and SRC-3 (loading control is β-actin). Modified from Mukherjee et al. (2006b) (Copyright 2006, *Molecular and Cellular Biology*)

et al. 2006b) by crossing a PR$^{Cre/+}$ knockin mouse (Soyal et al. 2005) with a SRC-2$^{flox/flox}$ mouse in which exon 11 of the SRC-2 gene was floxed to facilitate Cre-mediated excision (Gehin et al. 2002); exon 11 encodes the receptor interacting domain (RID). Therefore, the resultant PR$^{Cre/+}$ SRC-2$^{flox/flox}$ bigenic is designed to postnatally ablate SRC-2 only in cell lineages that express the PR (Mukherjee et al. 2006b). The advantage of this genetic approach is that SRC-2's involvement in PR-dependent transcriptional programs can be directly studied at the whole-organism level in the adult without interference from other unrelated phenotypes resulting from SRC-2's absence.

5 Uterine Receptivity Requires SRC-2 Function

As expected, female and male PR$^{Cre/+}$SRC-2$^{flox/flox}$ mice exhibit normal postnatal development; however, the PR$^{Cre/+}$SRC-2$^{flox/flox}$ female is infertile (Mukherjee et al. 2006b). Unlike the TIF2$^{-/-}$ model, male

In the case of the human endometrium, recent immunohistochemical studies clearly show that SRC-2 and PR are expressed in many of the same cell-types of the stromal and epithelial compartments (Fig. 4C–D); similar observations have been reported for the mouse (Mukherjee et al. 2006b).

As reported for the murine mammary gland (Mukherjee et al. 2006b), recent immunohistochemical studies of the normal human breast have revealed that a subset of epithelial cells score positive for SRC-2 expression (Fig. 4E and F). Interestingly, the restricted heterogeneous spatial expression pattern for SRC-2 in the human breast draws comparisons with a similar regional expression pattern for PR in rodent and human breast (Clarke et al. 1997), whether [like the mouse (Mukherjee et al. 2006b)] SRC-2 and PR colocalize in the human breast remains to be addressed. Further examination of this issue is important as segregation of mammary epithelial cells that express PR from cells that proliferate in response to progestin exposure is now accepted as an evolutionarily conserved feature that underlies a proposed paracrine mode of action ascribed to PR in the normal breast (reviewed in Fernandez-Valdivia et al. 2005).

8   Conclusions and Perspective

With over 200 known coactivators documented to date (Lonard and O'Malley 2006), it is remarkable that PR relies heavily on the coregulator properties of SRC-2 for a subset of progestin-induced physiological responses that are required for the maintenance of female fecundity and mammary morphogenesis in the mouse.

Markedly distinct from SRC-1 and SRC-3, whose coactivator functions serve a subset of progestin-initiated transcriptional programs either in the uterus or mammary gland, SRC-2 has been shown to be an essential PR coactivator in both target tissues. From a clinical perspective, the importance of SRC-2 in peri-implantation biology warrants further investigation, since recurrent implantation failure is now considered a key factor that undermines the establishment of a successful pregnancy [via natural means or by assisted reproductive technologies (ART) (Norwitz et al. 2001)]. Although the data are preliminary, abnormal elevations in SRC-2 levels have been detected in endometrial biopsies from infertile women with polycystic ovarian syndrome (PCOS) as well as in a subset of endometrial cancers (Gregory et al. 2002; Pathirage et al. 2006), suggesting a possible link between this coactivator and the pathogenesis of these uterine disorders. Considering the common use of progestins in the management of female reproductive health, coupled with our recent finding that SRC-2 is as an essential coactivator for PR action in female reproduction, SRC-2 may well represent a future molecular marker in diagnostic reproductive medicine or an intervention target for the treatment of a subset of gynecologic pathologies.

Although not required for the establishment of the mammary cell-lineage in which it is expressed (Ismail et al. 2002), the PR is essential for mammary epithelial expansion in response to progestin exposure (Lydon et al. 1995). The progestin-initiated proliferative signal represents a prerequisite step toward terminal differentiation of the mammary gland; importantly, however, this signal can also influence breast cancer susceptibility (reviewed in Fernandez-Valdivia et al. 2005). The observation that abrogation of SRC-2 in the murine mammary gland results in a near phenocopy of the PRKO mammary defect raises two important (and interconnected) questions: Can overexpression of SRC-2 promote neoplastic transformation in the murine mammary gland? And if so, does SRC-2 have a role in hormone-dependent breast cancers in the human? The answers to these questions promise to broaden our mechanistic understanding of progestin's role in breast cancer progression and may assist in the future design of more powerful diagnostic, prognostic, and/or therapeutic approaches in the clinical management of this disease.

In conclusion, progress in our mechanistic understanding of how different target tissues exhibit different responses to the same progestin will depend on identifying the key coregulators that are preferentially recruited by PR in a given target tissue. Identification of such coregulators will not only expand our current concepts on tissue-selective responses to progestins but may aid in the development of more efficacious progestin treatment strategies in the future. Therefore, we believe that continued studies of SRC-2 action in the human and mouse are required to further clarify this understudied yet important area of progestin research.

Acknowledgements. The technical assistance of Jie Li, Yan Ying, Jie Han, and Khauh Dao is gratefully acknowledged. We thank Drs. Pierre Chambon and Martine Gehin, Institut Clinique de la Souris, (ICS-IGBMC), BP10142, 67404 ILLKIRCH Cedex France, for providing the floxed SRC-2 mouse. This research was supported by NIH and private grants HD-42311 (F.J.D) and CA-77530 and Susan G. Komen Breast Cancer Research Program (J.P.L.).

References

Anzick SL, Kononen J, Walker RL, Azorsa DO, Tanner MM, Guan XY, Sauter G, Kallioniemi OP, Trent JM, Meltzer PS (1997) AIB1, a steroid receptor coactivator amplified in breast and ovarian cancer. Science 277:965–968

Belandia B, Parker MG (2000) Functional interaction between the p160 coactivator proteins and the transcriptional enhancer factor family of transcription factors. J Biol Chem 275:30801–30805

Beral V (2003) Breast cancer and hormone-replacement therapy in the Million Women Study. Lancet 362:419–427

Berrevoets CA, Umar A, Trapman J, Brinkmann AO (2004) Differential modulation of androgen receptor transcriptional activity by the nuclear receptor co-repressor (N-CoR). Biochem J 379:731–738

Brisken C, Heineman A, Chavarria T, Elenbaas B, Tan J, Dey SK, McMahon JA, McMahon AP, Weinberg RA (2000) Essential function of Wnt-4 in mammary gland development downstream of progesterone signaling. Genes Dev 14:650–654

Chatterton RT Jr, Lydon JP, Mehta RG, Mateo ET, Pletz A, Jordan VC (2002) Role of the progesterone receptor (PR) in susceptibility of mouse mammary gland to 7,12-dimethylbenz[a]anthracene-induced hormone-independent preneoplastic lesions in vitro. Cancer Lett 188:47–52

Chen D, Ma H, Hong H, Koh SS, Huang SM, Schurter BT, Aswad DW, Stallcup MR (1999) Regulation of transcription by a protein methyltransferase. Science 284:2174–2177

Chen D, Huang SM, Stallcup MR (2000a) Synergistic, p160 coactivator-dependent enhancement of estrogen receptor function by CARM1 and p300. J Biol Chem 275:40810–40816

Chen SL, Dowhan DH, Hosking BM, Muscat GE (2000b) The steroid receptor coactivator, GRIP-1, is necessary for MEF-2C-dependent gene expression and skeletal muscle differentiation. Genes Dev 14:1209–1228

Chen YH, Kim JH, Stallcup MR (2005) GAC63, a GRIP1-dependent nuclear receptor coactivator. Mol Cell Biol 25:5965–5972

Cheon YP, Li Q, Xu X, DeMayo FJ, Bagchi IC, Bagchi MK (2002) A genomic approach to identify novel progesterone receptor regulated pathways in the uterus during implantation. Mol Endocrinol 16:2853–2871

Clarke CA, Glaser SL, Uratsu CS, Selby JV, Kushi LH, Herrinton LJ (2006) Recent declines in hormone therapy utilization and breast cancer incidence: clinical and population-based evidence. J Clin Oncol 24:e49–50

Clarke RB, Howell A, Potten CS, Anderson E (1997) Dissociation between steroid receptor expression and cell proliferation in the human breast. Cancer Res 57:4987–4991

Colditz GA, Egan KM, Stampfer MJ (1993) Hormone replacement therapy and risk of breast cancer: results from epidemiologic studies. Am J Obstet Gynecol 168:1473–1480

Coste A, Antal MC, Chan S, Kastner P, Mark M, O'Malley BW, Auwerx J (2006) Absence of the steroid receptor coactivator-3 induces B-cell lymphoma. EMBO J 25:2453–2464

Culig Z, Klocker H, Bartsch G, Hobisch A (2002) Androgen receptors in prostate cancer. Endocr Relat Cancer 9:155–170

Das SK, Chakraborty I, Paria BC, Wang XN, Plowman G, Dey SK (1995) Amphiregulin is an implantation-specific and progesterone-regulated gene in the mouse uterus. Mol Endocrinol 9:691–705

Duong BN, Elliott S, Frigo DE, Melnik LI, Vanhoy L, Tomchuck S, Lebeau HP, David O, Beckman BS, Alam J, Bratton MR, McLachlan JA, Burow ME (2006) AKT regulation of estrogen receptor-beta transcriptional activity in breast cancer. Cancer Res 66:8373–8381

Duret L, Gasteiger E, Perriere G (1996) LALNVIEW: a graphical viewer for pairwise sequence alignments. Comput Appl Biosci 12:507–510

Fernandez-Valdivia R, Mukherjee A, Mulac-Jericevic B, Conneely OM, DeMayo FJ, Amato P, Lydon JP (2005) Revealing progesterone's role in uterine and mammary gland biology: insights from the mouse. Semin Reprod Med 23:22–37

Frigo DE, Basu A, Nierth-Simpson EN, Weldon CB, Dugan CM, Elliott S, Collins-Burow BM, Salvo VA, Zhu Y, Melnik LI, Lopez GN, Kushner PJ, Curiel TJ, Rowan BG, McLachlan JA, Burow ME (2006) p38 mitogen-activated protein kinase stimulates estrogen-mediated transcription and proliferation through the phosphorylation and potentiation of the p160 coactivator glucocorticoid receptor-interacting protein 1. Mol Endocrinol 20:971–983

Gehin M, Mark M, Dennefeld C, Dierich A, Gronemeyer H, Chambon P (2002) The function of TIF2/GRIP1 in mouse reproduction is distinct from those of SRC-1 and p/CIP. Mol Cell Biol 22:5923–5937

Gregory CW, Wilson EM, Apparao KB, Lininger RA, Meyer WR, Kowalik A, Fritz MA, Lessey BA (2002) Steroid receptor coactivator expression throughout the menstrual cycle in normal and abnormal endometrium. J Clin Endocrinol Metab 87:2960–2966

Gu YZ, Hogenesch JB, Bradfield CA (2000) The PAS superfamily: sensors of environmental and developmental signals. Annu Rev Pharmacol Toxicol 40:519–561

Han SJ, Jeong J, Demayo FJ, Xu J, Tsai SY, Tsai MJ, O'Malley BW (2005) Dynamic cell type specificity of SRC-1 coactivator in modulating uterine progesterone receptor function in mice. Mol Cell Biol 25:8150–8165

Han SJ, DeMayo FJ, Xu J, Tsai SY, Tsai MJ, O'Malley BW (2006) Steroid receptor coactivator (SRC)-1 and SRC-3 differentially modulate tissue-specific activation functions of the progesterone receptor. Mol Endocrinol 20:45–55

Hofman K, Swinnen JV, Verhoeven G, Heyns W (2002) Coactivation of an endogenous progesterone receptor by TIF2 in COS-7 cells. Biochem Biophys Res Commun 295:469–474

Huang ZJ, Edery I, Rosbash M (1993) PAS is a dimerization domain common to Drosophila period and several transcription factors. Nature 364:259–262

Ismail PM, Li J, DeMayo FJ, O'Malley BW, Lydon JP (2002) A novel lacZ reporter mouse reveals complex regulation of the progesterone receptor promoter during mammary gland development. Mol Endocrinol 16:2475–2489

Ismail PM, DeMayo FJ, Amato P, Lydon JP (2004) Progesterone induction of calcitonin expression in the murine mammary gland. J Endocrinol 180:287–295

Jeong JW, Lee KY, Kwak I, White LD, Hilsenbeck SG, Lydon JP, DeMayo FJ (2005) Identification of murine uterine genes regulated in a ligand-dependent manner by the progesterone receptor. Endocrinology 146:3490–3505

Jeong JW, Kwak I, Lee KY, White LD, Wang XP, Brunicardi FC, O'Malley BW, DeMayo FJ (2006) The genomic analysis of the impact of steroid receptor coactivators ablation on hepatic metabolism. Mol Endocrinol 20:1138–1152

Kim JH, Stallcup MR (2004) Role of the coiled-coil coactivator (CoCoA) in aryl hydrocarbon receptor-mediated transcription. J Biol Chem 279:49842–49848

Kim JH, Li H, Stallcup MR (2003) CoCoA, a nuclear receptor coactivator which acts through an N-terminal activation domain of p160 coactivators. Mol Cell 12:1537–1549

Kino T, Ichijo T, Chrousos GP (2004) FLASH interacts with p160 coactivator subtypes and differentially suppresses transcriptional activity of steroid hormone receptors. J Steroid Biochem Mol Biol 92:357–363

Kuang SQ, Liao L, Zhang H, Lee AV, O'Malley BW, Xu J (2004) AIB1/SRC-3 deficiency affects insulin-like growth factor I signaling pathway and suppresses v-Ha-ras-induced breast cancer initiation and progression in mice. Cancer Res 64:1875–1885

Kuang SQ, Liao L, Wang S, Medina D, O'Malley BW, Xu J (2005) Mice lacking the amplified in breast cancer 1/steroid receptor coactivator-3 are resistant to chemical carcinogen-induced mammary tumorigenesis. Cancer Res 65:7993–8002

Lee S, Mohsin SK, Mao S, Hilsenbeck SG, Medina D, Allred DC (2005) Hormones, receptors, and growth in hyperplastic enlarged lobular units: early potential precursors of breast cancer. Breast Cancer Res 8:1–9

Lee YH, Stallcup MR (2006) Interplay of Fli-I and FLAP1 for regulation of beta-catenin dependent transcription. Nucleic Acids Res 34:5052–5059

Lim H, Ma L, Ma WG, Maas RL, Dey SK (1999) Hoxa-10 regulates uterine stromal cell responsiveness to progesterone during implantation and decidualization in the mouse. Mol Endocrinol 13:1005–1017

Lonard DM, O'Malley BW (2005) Expanding functional diversity of the coactivators. Trends Biochem Sci 30:126–132

Lonard DM, O'Malley BW (2006) The expanding cosmos of nuclear receptor coactivators. Cell 125:411–414

Lonard DM, Tsai SY, O'Malley BW (2004) Selective estrogen receptor modulators 4-hydroxytamoxifen and raloxifene impact the stability and function of SRC-1 and SRC-3 coactivator proteins. Mol Cell Biol 24:14–24

Louet JF, Coste A, Amazit L, Tannour-Louet M, Wu RC, Tsai SY, Tsai MJ, Auwerx J, O'Malley BW (2006) Oncogenic steroid receptor coactivator-3 is a key regulator of the white adipogenic program. Proc Natl Acad Sci U S A 103:17868–17873

Lydon JP, DeMayo FJ, Funk CR, Mani SK, Hughes AR, Montgomery CA Jr, Shyamala G, Conneely OM, O'Malley BW (1995) Mice lacking progesterone receptors exhibit pleiotropic reproductive abnormalities. Genes Dev 9:2266–2278

Lydon JP, Ge G, Kittrell FS, Medina D, O'Malley BW (1999) Murine mammary gland carcinogenesis is critically dependent on progesterone receptor function. Cancer Res 59:4276–4284

Mark M, Yoshida-Komiya H, Gehin M, Liao L, Tsai MJ, O'Malley BW, Chambon P, Xu J (2004) Partially redundant functions of SRC-1 and TIF2 in postnatal survival and male reproduction. Proc Natl Acad Sci U S A 101:4453–4458

McKenna NJ, O'Malley BW (2002) Combinatorial control of gene expression by nuclear receptors and coregulators. Cell 108:465–474

Medina D, Kittrell FS, Shepard A, Contreras A, Rosen JM, Lydon J (2003) Hormone dependence in premalignant mammary progression. Cancer Res 63:1067–1072

Modder UI, Sanyal A, Kearns AE, Sibonga JD, Nishihara E, Xu J, O'Malley BW, Ritman EL, Riggs BL, Spelsberg TC, Khosla S (2004) Effects of loss of steroid receptor coactivator-1 on the skeletal response to estrogen in mice. Endocrinology 145:913–921

Mukherjee A, Amato P, Allred DC, Fernandez-Valdivia R, Nguyen J, O'Malley BW, DeMayo FJ, Lydon JP (2006a) Steroid receptor coactivator 2 is essential for progesterone-dependent uterine function and mammary morphogenesis: insights from the mouse—implications for the human. J Steroid Biochem Mol Biol 102:22–31

Mukherjee A, Soyal SM, Fernandez-Valdivia R, Gehin M, Chambon P, DeMayo FJ, Lydon JP, O'Malley BW (2006b) Steroid receptor coactivator 2 is critical for progesterone-dependent uterine function and mammary morphogenesis in the mouse. Mol Cell Biol 26:6571–6583

Norwitz ER, Schust DJ, Fisher SJ (2001) Implantation and the survival of early pregnancy. N Engl J Med 345:1400–1408

Patchev AV, Fischer D, Wolf SS, Herkenham M, Gotz F, Gehin M, Chambon P, Patchev VK, Almeida OF (2007) Insidious adrenocortical insufficiency underlies neuroendocrine dysregulation in TIF-2 deficient mice. FASEB J 21:231–238

Pathirage N, Di Nezza LA, Salmonsen LA, Jobling T, Simpson ER, Clyne CD (2006) Expression of aromatase, estrogen receptors, and their coactivators in patients with endometrial cancer. Fertil Steril 86:469–472

Rogatsky I, Luecke HF, Leitman DC, Yamamoto KR (2002) Alternate surfaces of transcriptional coregulator GRIP1 function in different glucocorticoid receptor activation and repression contexts. Proc Natl Acad Sci U S A 99:16701–16706

Ross RK, Paganini-Hill A, Wan PC, Pike MC (2000) Effect of hormone replacement therapy on breast cancer risk: estrogen versus estrogen plus progestin. J Natl Cancer Inst 92:328–332

Rossouw JE, Anderson GL, Prentice RL, LaCroix AZ, Kooperberg C, Stefanick ML, Jackson RD, Beresford SA, Howard BV, Johnson KC, Kotchen JM, Ockene J (2002) Risks and benefits of estrogen plus progestin in healthy postmenopausal women: principal results from the Women's Health Initiative randomized controlled trial. JAMA 288:321–333

Schairer C, Lubin J, Troisi R, Sturgeon S, Brinton L, Hoover R (2000) Menopausal estrogen and estrogen-progestin replacement therapy and breast cancer risk. JAMA 283:485–491

Soyal SM, Mukherjee A, Lee KY, Li J, Li H, DeMayo FJ, Lydon JP (2005) Cre-mediated recombination in cell lineages that express the progesterone receptor. Genesis 41:58–66

Takamoto N, Zhao B, Tsai SY, DeMayo FJ (2002) Identification of Indian hedgehog as a progesterone-responsive gene in the murine uterus. Mol Endocrinol 16:2338–2348

Torres-Arzayus MI, Font de Mora J, Yuan J, Vazquez F, Bronson R, Rue M, Sellers WR, Brown M (2004) High tumor incidence and activation of the PI3K/AKT pathway in transgenic mice define AIB1 as an oncogene. Cancer Cell 6:263–274

Wang Z, Rose DW, Hermanson O, Liu F, Herman T, Wu W, Szeto D, Gleiberman A, Krones A, Pratt K, Rosenfeld R, Glass CK, Rosenfeld MG (2000) Regulation of somatic growth by the p160 coactivator p/CIP. Proc Natl Acad Sci U S A 97:13549–13554

Wang Z, Qi C, Krones A, Woodring P, Zhu X, Reddy JK, Evans RM, Rosenfeld MG, Hunter T (2006) Critical roles of the p160 transcriptional coactivators p/CIP and SRC-1 in energy balance. Cell Metab 3:111–122

Xu J, Qiu Y, DeMayo FJ, Tsai SY, Tsai MJ, O'Malley BW (1998) Partial hormone resistance in mice with disruption of the steroid receptor coactivator-1 (SRC-1) gene. Science 279:1922–1925

Xu J, Liao L, Ning G, Yoshida-Komiya H, Deng C, O'Malley BW (2000) The steroid receptor coactivator SRC-3 (p/CIP/RAC3/AIB1/ACTR/TRAM-1) is required for normal growth, puberty, female reproductive function, and mammary gland development. Proc Natl Acad Sci U S A 97:6379–6384

Ye X, Han SJ, Tsai SY, DeMayo FJ, Xu J, Tsai MJ, O'Malley BW (2005) Roles of steroid receptor coactivator (SRC)-1 and transcriptional intermediary factor (TIF) 2 in androgen receptor activity in mice. Proc Natl Acad Sci U S A 102:9487–9492

Ying H, Furuya F, Willingham MC, Xu J, O'Malley BW, Cheng SY (2005) Dual functions of the steroid hormone receptor coactivator 3 in modulating resistance to thyroid hormone. Mol Cell Biol 25:7687–7695

Zhou HJ, Yan J, Luo W, Ayala G, Lin SH, Erdem H, Ittmann M, Tsai SY, Tsai MJ (2005) SRC-3 is required for prostate cancer cell proliferation and survival. Cancer Res 65:7976–7983

Ernst Schering Foundation Symposium Proceedings, Vol. 1, pp. 77–107
DOI 10.1007/2789_2008_076
© Springer-Verlag Berlin Heidelberg
Published Online: 29 February 2008

Progesterone Receptor Isoforms in Normal and Malignant Breast

P.A. Mote, J.D. Graham, C.L. Clarke(✉)

Westmead Institute for Cancer Research, University of Sydney at Westmead Millennium
Institute, Westmead Hospital, Darcy Rd, 2145 Westmead, Australia
email: *christine_clarke@wmi.usyd.edu.au*

Abstract. Progesterone is an essential regulator of normal female reproductive function, yet recent studies on the use of progestins in hormone replacement therapy have clearly implicated progestins in breast cancer development, a disease initiated early in life at a time of frequent exposure to cycling ovarian hormones. The effects of progesterone are mediated by two distinct nuclear receptor proteins, PRA and PRB. In normal breast PRA and PRB are co-expressed at similar levels in luminal epithelial cells, suggesting that both proteins are required to mediate physiologically relevant progesterone signalling. However, early in breast carcinogenesis PRA:PRB expression is disrupted, resulting in frequent predominance of one isoform. Unbalanced expression of PRA and PRB results in altered hormonal response and aberrant targeting of genes that are not normally progestin-regulated, principally those involved in morphological changes and disruptions of the actin cytoskeleton, and in migration. Movement of PR into discrete nuclear domains, or foci, is a critical step in normal PR transcriptional activity that appears to be aberrant in cancers and likely related to alterations in nuclear morphology, gene expression and cell function associated with tumour cells. Given that exogenous progestins are consumed by millions of women worldwide in the form of hormone replacement therapy and oral contraceptives, it is vital to better understand the mechanisms of progesterone action in the breast.

1   Introduction

Progesterone is an important physiological modulator that underpins normal reproductive function and provides significant benefits to many women worldwide. Healthy women are regularly exposed to progestational agents (progestins) in contraceptive formulations and in hormone replacement therapy (HRT) and progesterone is a common treatment for cycle-related breast and uterine disorders. Progesterone plays a central role in normal female reproduction, in the uterus, ovary, mammary gland and brain, and the number of cellular pathways known to be regulated by progesterone indicates the complexity of its physiological role (Graham and Clarke 1997). Although progesterone is one of the key ovarian regulators of reproductive function in normal physiology, exogenous exposure to progestins increases breast cancer risk (Rossouw et al. 2002; Beral 2003; Santen 2003; Lee et al. 2005).

1.1   Progesterone Action in the Human Breast

The mature breast is composed of a number of ductular trees that drain blind-ended glandular clusters (lobules) and exit the nipple. Histologically the ducts and lobules are lined by a single layer of epithelial cells that are surrounded by myoepithelial cells. The lobules are embedded in loose connective tissue known as intralobular stroma that contains stromal and other cells, as distinct from the denser interlobular stroma, found elsewhere in the breast. The systemic endocrine environment is active during the major developmental stages of the breast, implicating ovarian hormones in this process (Howard and Gusterson 2000), and animal studies have provided further support for the essential roles of oestrogen and progesterone in mammary gland development and function (Bocchinfuso and Korach 1997; Conneely et al. 2001).

Breast development occurs primarily at puberty through rapid expansion of the tissue under the influence of ovarian hormones. Progesterone is one of the master regulators that controls distinct aspects of this programme, particularly ductal side-branching and lobular development (Conneely et al. 2001). Progesterone regulates the normal breast during the menstrual cycle, as evidenced by breast expansion as a consequence of enlargement of the epithelial component (Soderqvist et al. 1997), and stromal oedema and vacuolisation of the myoepithelium also occur in the progesterone-dominated luteal phase (Longacre and Bartow 1986). In pregnancy, progesterone regulates further development of the breast, whilst withdrawal of progesterone at the end of pregnancy facilitates parturition and lactation (Topper and Freeman 1980).

Oestrogen and progesterone mediate their effects via specific nuclear receptors, ER and PR, which are expressed exclusively in luminal epithelial cells in the normal breast. Receptor positivity is observed in 20–30% of luminal epithelial cells (Anderson 2002) and current evidence suggests that receptor-positive cells do not respond directly to hormone signal, but rather exert a paracrine influence on the biology of surrounding receptor-negative cells (Brisken et al. 1998). PR is expressed as two proteins, PRA and PRB, and there is increasing evidence that the isoforms are functionally different (Shyamala et al. 1998; Mulac-Jericevic et al. 2000; Graham and Clarke 2002; Mulac-Jericevic et al. 2003). Generally PRA and PRB are co-expressed in the same target cells in human tissues, and their relative expression, where it has been examined, is usually close to unity (Mote et al. 1999, 2002). However, in some normal physiological circumstances and in some cell types there is predominance of one isoform. For example, in the normal uterine stroma PRA is always the predominant isoform, and in the epithelial glands PRB is predominantly expressed during the mid-secretory phase of the cycle (Mote et al. 1999). In addition, whilst in the majority of non-human species both PRA and PRB are expressed, the relative expression of the two proteins can diverge from unity. PRA is always the predominant isoform in the rodent (Schneider et al. 1991) and is widely expressed in the macaque reproductive system (Isaksson et al. 2003).

1.2 Transcriptional Regulation by PR

PR is a member of a large family of ligand-activated nuclear transcription regulators that are characterised by organisation into specific functional domains and are conserved, to varying degrees, between species and family members (Germain et al. 2003). PR comprises a central DNA binding domain and a carboxyl-terminal ligand-binding domain. In addition, the receptor contains a number of activation (AF) and inhibitory (IF) function elements which enhance and repress transcriptional activation of PR by association of these regions with transcriptional co-regulators (McKenna and O'Malley 2002). The expression of human PR is controlled by two promoters which direct the synthesis of two distinct subgroups of mRNA transcripts (Kastner et al. 1990) encoding the two receptor proteins. The two PR forms are identical except that PRA lacks 164 amino acids of the N-terminal end of PRB. The region of the protein that is unique to PRB contains a transcription activation function, AF3 (Sartorius et al. 1994), in addition to AF1 and AF2, which are common to PRA. Newly transcribed cytoplasmic PR is assembled in an inactive multi-protein chaperone complex which dissociates upon ligand binding and receptor activation. Progestin binding to PR causes a conformational change and dimerisation, resulting in association of the progestin-complexed PR dimer with specific co-activators and general transcription factors and binding to progestin response elements (PREs) in the promoters of target genes. This results in the transcriptional regulation of those genes (reviewed in Tsai and O'Malley 1994; McKenna and O'Malley 2002).

When the transcriptional activities of PRA and PRB are examined, by transient co-transfection of PRA and/or PRB and reporter constructs containing progestin-responsive sequences into a variety of cell lines, there is considerable evidence of differences in the transcriptional activities of PRA and PRB. In all cell types examined PRB exhibits hormone-dependent transactivation irrespective of the complexity of the response elements, whereas the transcriptional activity of PRA is cell- and reporter-specific (Tung et al. 1993; Vegeto et al. 1993; Sartorius et al. 1994). Furthermore, PRA acts as a transdominant inhibitor of PRB in situations where PRA has little or no transactivational activity (Tung et al. 1993; Vegeto et al. 1993). PRA can also regulate the transcriptional activity of other nuclear receptors such as glucocorticoid, mineralocorticoid, androgen and oestrogen (Wiehle et al. 1990; Tung et al. 1993; Vegeto et al. 1993; McDonnell and Goldman 1994; Wen et al. 1994), suggesting that this isoform may play a central role in regulating the activity of a number of nuclear receptors in addition to PRB. However, the ability of PRA to act as a transdominant repressor is very much model-specific, and there is considerable contradiction between reports. In support of the distinct activities of the two PR proteins, patterns of gene regulation in T-47D breast cancer cells expressing only PRA or PRB reveal a remarkably small overlap between the genes regulated by the two receptors, with the subset of genes regulated by PRB greatly exceeding in number those regulated by PRA (Richer et al. 2002).

The mechanisms by which PRA and PRB exert such apparently different transcriptional activities in various cell and promoter experimental systems remain for the most part unknown, although a number of possible scenarios have been suggested. The physical differences at the N-terminal end of the two receptors are liable to be responsible for some transcriptional differences. In addition to the fact that AF3 is unique to PRB, the B-specific region has a distinct conformation in solution (Bain et al. 2001) that is likely to obscure an inhibitory domain active in the N-terminus of the PRA protein (Huse et al. 1998). The unique AF in PRB may also confer differing co-regulator affinities between the two

PR proteins (Giangrande et al. 2000) and, given that PR acts in combination with multiple other transcription factors to affect transcription, it is possible that variability of the tissue-specific expression of the components of this multi-protein complex would result in different PRA and PRB activities within the same cell.

Although when individually examined PRA and PRB exhibit different transcriptional activities, with PRA often acting as a dominant inhibitor of PRB activity, the experimental scenarios in which these data were obtained do not mimic the physiological situation. In normal tissues PRA and PRB are co-expressed in target cells, and in breast cancer, although there are disrupted ratios of PRA:PRB, both PR isoforms can always be detected (Mote et al. 2002). Based on the experimental studies cited above, it would be expected, if the two isoforms are transcriptionally distinct, that changes in the relative amounts of PRA and PRB such as are observed in breast cancer would result in altered target gene expression patterns and moreover that predominance of PRA would be inhibitory. However, this is not the case. In PR-positive T-47D breast cancer cells co-expressing PRA and PRB at similar levels, there was a remarkably small overlap between the genes regulated by progestins in these cells (Graham et al. 2005) and in cells that expressed either PRA or PRB but not both (Richer et al. 2002). Moreover, markedly fewer genes were progestin-regulated in cells expressing both PRA and PRB when compared with cells expressing only one PR protein (Graham et al. 2005). This is consistent with the view that a substantial proportion of progestin-regulated transcripts are targets of PRA or PRB homodimers, but not heterodimers, and that extrapolation of gene expression data from experimental systems that do not mimic the physiological context of co-expressed PRA and PRB needs to be approached with caution.

To examine the dominant inhibitory effect of PRA observed in transfected systems the relative expression of PRA and PRB was varied in wild-type T-47D cells, which already express both isoforms. The impact on transcription was not striking unless PRA was in vast excess over PRB (McGowan et al. 2003). Progestin regulation profiles were very similar between cells expressing equivalent PRA and PRB levels, and cells in which PRA was the predominant species (Graham et al. 2005). There was no evidence of dominant transcriptional inhibition by

PRA, suggesting that this phenomenon may only be observed in transient transfection models where much greater excesses of PRA expression can be achieved.

1.3   Endocrine Signalling and Breast Cancer Development

Ovarian activity is implicated in modulating the risk of breast cancer. Breast carcinoma arises from the ductal epithelium, and ovarian hormones are implicated in its full development: women without ovaries have a dramatically reduced cancer risk (Santen et al. 2001), and moreover a shorter reproductive life, during which time a woman is exposed less to endogenous cycling oestrogen and progesterone, is known to be protective (Clavel-Chapelon and Gerber 2002). Controlling the endocrine environment can also reduce breast cancer risk, with successful preventive endocrine intervention strategies using anti-oestrogens and aromatase inhibitors for vulnerable women (Yue et al. 2005). The majority of established breast cancers are hormone-responsive and manipulation of the endocrine environment is a cornerstone of current treatment, in the form of adjuvant tamoxifen or aromatase therapy. Endocrine manipulation to inhibit the availability of oestrogen is an important form of adjuvant treatment for women with ER-positive breast cancer (Dowsett et al. 2005).

Changes to the hormonal milieu frequently occur in the breast of healthy women through use of exogenous hormones, such as HRT and oral contraception (OCP), and it is well established that there is a link between their use and elevated breast cancer risk (Rossouw et al. 2002; Beral 2003; Santen 2003; Lee et al. 2005), although the mechanisms involved remain unclear. Women exposed to progestin-containing HRT have an increased breast cancer risk compared to women taking oestrogen alone (Beral 2003), and HRT with high progestin potency increases mammographic density (Greendale et al. 2003; McTiernan et al. 2005), which is associated with increased breast cancer risk. HRT is consumed by millions of women worldwide, highlighting the need to better understand both how progesterone acts in the normal breast and how it increases breast cancer risk.

1.4   Exposure of Early Breast Cancers to Cyclical Reproductive Hormones

Invasive cancer is most commonly diagnosed in women over 50, and ovarian hormones are thought to act as promoters in cells with genetic changes. There is now considerable evidence that genetically altered ducts/lobules are scattered within the normal breast (Larson et al. 1998; Gong et al. 2001; Larson et al. 2002). The fact that genetic changes are detectable by molecular analysis indicates that such areas contain significant numbers of altered cells, supporting the view that these have arisen by proliferation of genetically altered clones (Smalley and Ashworth 2003).

The emergence of breast cancer in later life is consistent with the accepted multi-step model of carcinogenesis, and recent profiling evidence suggests that breast cancers are grouped into a number of distinct subtypes (Sorlie et al. 2001), based on cell lineage composition, and supporting the view that there are different cells of origin and/or a different sequence and nature of genetic alterations. The character of the initial genetic changes remains largely undefined, except in the case of the less than 10% of breast cancers that are familial. Germline mutations in one of the BRCA susceptibility genes are associated with markedly increased breast cancer risk (Venkitaraman 2002), demonstrating that mutations in these genes are critical initiating events in some women. In the majority of breast cancers, however, environmental factors such as ionising radiation (Park et al. 2003) or exposure to carcinogens (Brody et al. 2007) are likely initial triggers. As continued replication is required to progress the 'malignant cascade', cancers probably originate in a progenitor cell that has received a carcinogenic hit. These affected progenitor cells proliferate and form a larger population of epithelial cells with genetic mutations. This results in a 'field' of affected cells (Garcia et al. 1999), increasing the target population susceptible to receiving further hits stochastically.

The initial genetic hits probably occur in puberty, when the most rapid expansion of the breast tissue takes place, and therefore cells with genetic changes, which are likely to remain morphologically undetectable for a number of years, will be exposed for an extended period to regular cycling hormones including oestrogen and progesterone. The role of progesterone in breast cancer development is less well established than that of oestrogen, but animal studies show that PR is required for mammary carcinogenesis (Lydon et al. 1999) and there is evidence that progesterone can increase genomic instability in morphologically normal cells with a genetic change (Goepfert et al. 2000), likely to be due in part to regulation of centrosome function (Goepfert et al. 2002).

2   PRA and PRB Co-expression in Breast Tissues

A number of years ago our laboratory developed a dual immunofluorescent technique to reveal the levels of PRA and PRB expression simultaneously within the same archival tissue section (Mote et al. 1999). This allowed visualisation and measurement of the individual PR isoforms so that their relative expression levels could be determined and changes in the PRA:PRB ratio could be evaluated. We have demonstrated in vivo, in both breast and uterus, that PRA and PRB are co-expressed in normal human target cells and that all PR-positive epithelial cells co-express PRA and PRB at similar levels (Mote et al. 1999, 2000, 2002). This suggests that both PR proteins are required to mediate physiologically relevant progesterone signalling. Moreover, the heterodimer is the likely active species in the human, as we have shown that the two proteins co-locate to nuclear foci in normal tissues in vivo (Arnett-Mansfield et al. 2004, 2007).

Whilst PRA and PRB were co-expressed, at similar levels, in the normal breast, carcinogenesis was frequently accompanied by alterations in relative PRA and PRB expression (Mote et al. 2002) that could be both functionally and clinically important. Our studies suggested that disrupted PR protein expression is likely to be an intrinsic feature of cancer, as it was noted early during disease progression in atypical proliferative lesions.

2.1   Unbalanced PRA and PRB Expression in Normal Breast of Women at High Risk of Breast Cancer

A large proportion of familial breast and ovarian cancers are due to the breast cancer susceptibility genes *BRCA1* and *BRCA2*, although it remains unclear how disruptions in the functions of these proteins can increase cancer risk preferentially in hormone-dependent tissues. Women with a mutation in *BRCA1* or *BRCA2* have a significantly higher risk of developing breast cancer compared to the general population (Venkitaraman 2002). By measuring hormone responsiveness in prophylactically removed normal breast tissues from women with a germline pathogenic mutation in one of the *BRCA* genes we have demonstrated that expression of PR is significantly altered (Mote et al. 2004). Compared to control cases, in which the PRA and PRB isoforms were equally expressed, in mutation carriers the PRB protein was notably absent, resulting in predominance of the PRA isoform in these tissues (Fig. 1). An imbalance in the relative levels of PRA and PRB is likely to result in changes in progesterone signalling in hormone-dependent tissues, and may contribute to an altered hormonal milieu able to facilitate subsequent events in the development of breast cancer.

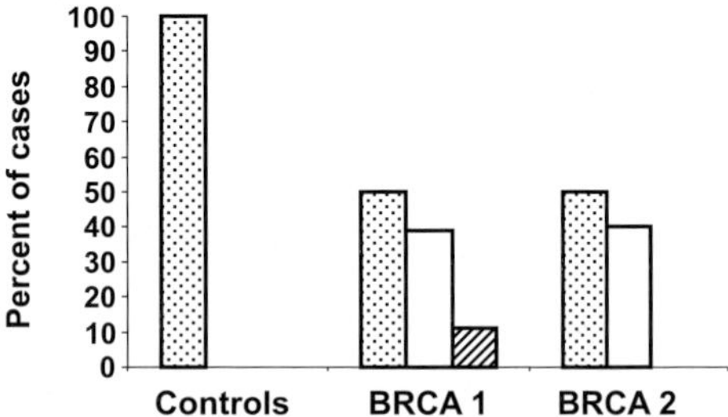

Fig. 1. Expression of PR isoforms in normal breast tissues of women with a germline pathogenic mutation in one of the BRCA genes. Sections were stained for PRA or PRB by immunohistochemistry and scored by three independent observers. Percentage of cases in each cohort that: express both PRA and PRB proteins (*stippled bars*), express PRA proteins only (*open bars*), express PRB proteins only (*hatched bars*). (Figure reproduced from Mote et al. 2004, with permission of John Wiley & Sons, Inc.)

2.2   Disrupted PRA and PRB Expression in Premalignant Breast Tissue

Progression from the normal state to malignancy in the human breast was accompanied by a progressively unequal expression of PRA and PRB that was often associated with marked adjacent cell differences (heterogeneity) in the relative expression of PR isoforms (Fig. 2). This was in striking contrast to the homogeneity and equivalent expression of PRA and PRB observed in normal tissues (Mote et al. 2002). Disruption of PR isoform expression appeared early in the development of breast cancer and was detectable in atypical hyperplasias (ADH), although it was not usually observed in proliferative disease without atypia (PDWA) (Mote et al. 2002). Disruption may lead to inappropriate signalling by ovarian hormones resulting in aberrant proliferative, invasive potential, and subsequent tumour formation and/or growth.

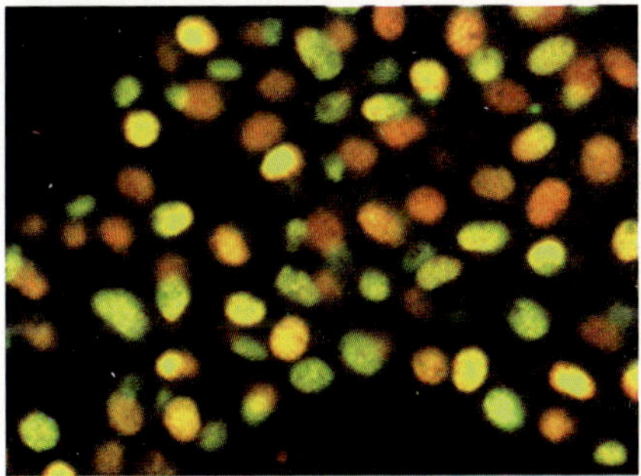

Fig. 2. Heterogeneous cell-to-cell expression of PRA and PRB. Formalin-fixed paraffin-embedded tissues were stained by dual immunofluorescence to demonstrate PRA and PRB proteins individually within the same section when viewed under dual immunofluorescent excitation. Cells expressing a predominance of PRA or PRB appear *green* or *orange* respectively; cells expressing equivalent levels of *PRA* and *PRB* are yellow. Original magnification ×400

2.3 Loss of Concordant PR Isoform Expression in Breast Cancers

There was significant disruption of PRA:PRB expression associated with breast cancer; although most tumours continued to express both PR isoforms, marked predominance of one isoform, especially PRA, was common (Mote et al. 2002). This supported our earlier immunoblot studies in which PRA-predominant tumours were significantly over represented (Graham et al. 1995). More recently alterations in PRA:PRB expression have been linked to treatment response, and the over-expression of PRA is associated with resistance to tamoxifen (Hopp et al. 2004; Osborne et al. 2005). From a clinical perspective, this suggests that the relative levels of PRA and PRB may be informative in the initial choice of treatment for the individual patient.

Using dual immunofluorescence we examined variability in relative levels of PRA and PRB in a cohort of PR-positive primary breast cancers (Fig. 3). Cases were assigned into one of three categories (PRA= PRB; PRA>PRB; PRB>PRA) based on image analyses of the levels of expression of the individual PR proteins. We reasoned that small deviations from PRA:PRB ratios of 1 were unlikely to have biological significance, and accordingly the scoring method ensured that only cases with a clear predominance of one isoform were scored as being PRA>PRB or PRB>PRA. We found that more than half of the tumours retained similar levels of PRA and PRB (Fig. 3a), whereas just over one-third showed a significant over-expression of the PRA protein (Fig. 3b) and around 10% a predominance of PRB (Fig. 3c) (Mote et al. 2002).

Predominant expression of PRA or PRB could result either from a marked increase in expression of one isoform without concomitant loss of the other, or alternatively from loss of one PR isoform. However, there appeared to be no association between the overall level of PR expression and the relative levels of PRA and PRB (P.A. Mote, A. Gompel, L.R. Webster, A. Lavaur, Y. Decroix, D. Hugol, R.L. Ward, N.J. Hawkins, K. Byth, R.L. Balleine, C.L. Clarke, unpublished observations), suggesting that predominance of PRA or PRB was likely to be due to an alteration in the balance of the levels of the two isoforms, rather than to a preferential loss of one isoform.

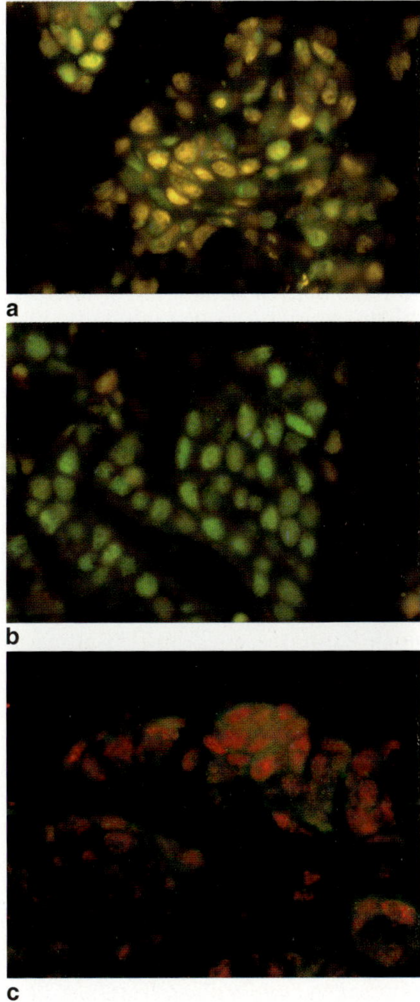

Fig. 3a–c. Expression of PR isoforms in invasive breast lesions showing **a** similar levels of PRA and PRB, **b** predominant expression of PRA and **c** predominant expression of PRB. Expression and relative levels of PRA and PRB were determined by dual immunofluorescent histochemistry. Original magnification ×400

The finding that cancer tissues had relative levels of PRA and PRB widely divergent from those seen in normal cells, and that predominance of one isoform was common, points to the probability that the homodimer is the active PR dimer species in cancers, unlike the heterodimer in normal tissues.

3   Altered PRA/B Ratio Has Physiological Consequences

Evidence from in vitro transcriptional studies and from transgenic and knockout animal models demonstrates that PRA and PRB are functionally different. This suggests that an altered balance of the two PR isoforms in hormone responsive tissues is likely to have biological and physiological consequences.

3.1   Genes Involved in Cell Shape and Adhesion Acquire Progesterone Responsiveness When PRA Exceeds PRB

The disruption of PR isoform expression, resulting in PRA predominance, is frequently observed in primary breast cancers, and in vitro studies of progestin effects on aspects of actin microfilament composition in PRA-predominant breast cancer cells provide evidence for endocrine regulation of changes in the cytoskeleton known to be associated with cancer development (McGowan et al. 2003). Moreover, by the use of cell lines over-expressing PRA we demonstrated that whilst the majority of PR-regulated genes were not sensitive to increased PRA: PRB ratio, a small but important subgroup of specific gene targets acquired progestin responsiveness (Graham et al. 2005). This small proportion (14%) of genes either acquired or lost progestin regulation with a high PRA:PRB ratio, and most acquired responsiveness to progestin in signalling pathways associated with cell shape, adhesion and survival. Given that PR isoform disruption occurs early in malignant progression these data have important implications for breast cancer biology; such a disruption may contribute to altered regulation of morphology and subsequent tumour development.

3.2 Increased PRA:PRB Ratio Leads to Altered Cytoskeletal and Adhesion Signalling in Breast Cancer Cells

The established role of progesterone in directing normal mammary gland development demonstrates its capacity to regulate epithelial cell shape and adhesion. This was confirmed in PR-positive T-47D breast cancer cells, where treatment with progestins resulted in marked cell flattening (Fig. 4a(b)) due to reorganisation of the actin cytoskeleton and loss of focal adhesions (McGowan et al. 2003). This effect was dependent on equivalent expression of PRA and PRB. An imbalance of PR isoforms resulted in a dramatically different response in these cells. In our model of PRA predominance, progestin treatment of cells expressing PRA in fivefold excess over PRB resulted in marked cell rounding (McGowan et al. 2003). Moreover, in contrast to the larger, flattened cells observed after progestin exposure when PRA and PRB were equivalently expressed, cells became significantly smaller than vehicle-treated controls (Fig. 4a(d)) and detached easily from the tissue culture substrate (McGowan et al. 2003). The finding that the small subgroup of genes that became regulated by progestins in cells in which PRA was over-expressed was enriched for transcripts encoding proteins involved in adhesion and maintenance of cell shape was consistent with this observation. Indeed, when polymerised actin and the focal adhesion proteins ezrin or paxillin were visualised in these cells, stress fibres and focal adhesions were lost, both when PRA and PRB were equivalent and when PRA was predominant (McGowan et al. 2003; Graham et al. 2005). However, in PRA over-expressing cells the effect was aberrant: ezrin and polymerised actin became aggregated into pools in the cytoplasm (Fig. 4b), suggesting a failure of the dynamic turnover of these cytoskeletal components (McGowan et al. 2003). The aberrant cytoskeletal and focal adhesion protein regulation by progestins in cells with a high PRA:PRB ratio had functional consequences. In a model of breast cancer cell invasion, T-47D cells moved readily into a bone marrow fibroblast layer and this was inhibited by progestins (McGowan et al. 2004). Over-expression of PRA abrogated this inhibition, suggesting that breast cancers that express high levels of PRA may be refractory to the inhibitory influence of progesterone and may be able to invade the surrounding stroma more easily (Fig. 5). The progesterone-mediated

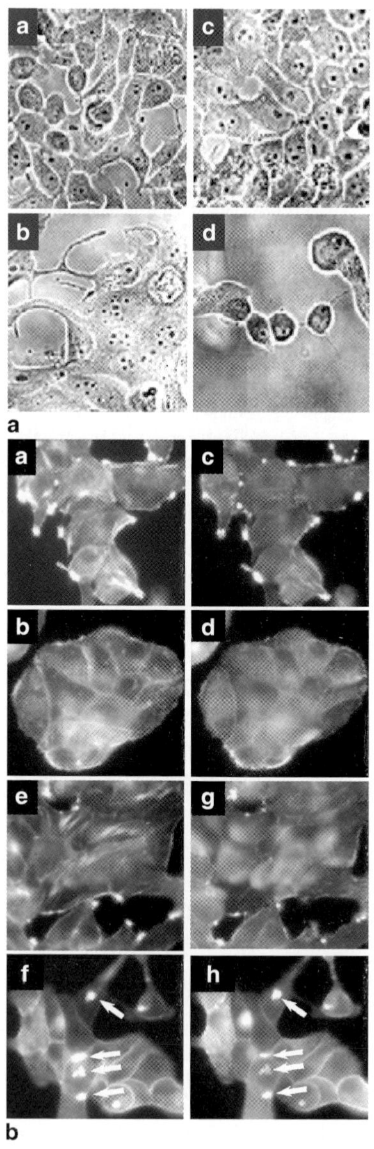

Fig. 4a, b. Progestin's effects on cell morphology in T-47DhPRA cells. **a** T-47DhPRA cells were seeded in 25 cm^2 flasks at a density of 2×10^4 cells/ml. Cells were treated with isopropyl-beta-D-thiogalactopyranoside (IPTG) (10 mM) or vehicle for 24 h then treated with ORG2058 (10 nM) or vehicle. Photographs *a–d* show the morphology of the cells 72 h later. (*a*) Vehicle, (*b*) ORG2058, (*c*) IPTG, (*d*) IPTG and ORG2058. **b** Cells were treated for 24 h with IPTG [10 mM (e–h)] or vehicle (*a–d*), then 48 h with ORG2058 (10 nM) or vehicle. Control (*a, c, e, g*) and ORG2058 (*b, d, f, h*) treated cells were fixed and stained, to visualise polymerised actin (*a, b, e, f*) and ezrin (*c, d, g, h*). *Arrows* indicate aggregates of ezrin and polymerised actin. (Figure redrawn from McGowan et al. 2003; Graham et al. 2005)

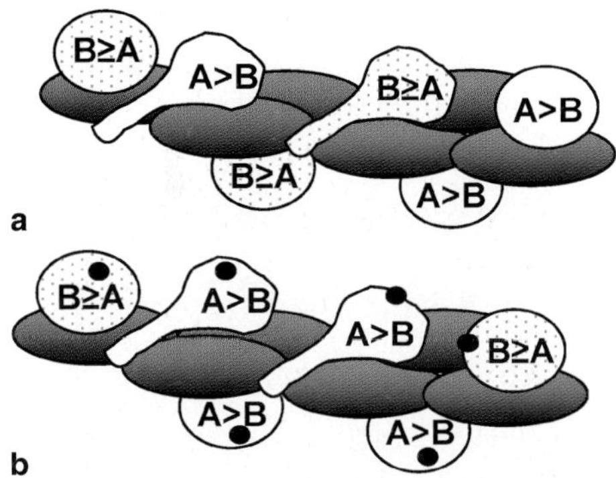

Fig. 5a, b. Breast cancer cell interaction with bone marrow fibroblasts. **a** Breast cancer cells migrate into fibroblast layers in the absence of progestin regardless of the PRA:PRB ratio. **b** In the presence of progestin (*black sphere*) migration of cells with PRB≥PRA (*stippled*) is inhibited whilst cells where PRA>PRB (*clear*) continue to invade stromal layers. (Figure reproduced, with kind permission of Springer Science and Business Media, from McGowan et al. 2004)

loss of focal adhesions inhibited the capacity of T-47D cells to re-adhere to the culture substrate if dislodged. This effect was significantly greater if PRA was predominant, suggesting that aberrant cytoskeletal dynamics in these cells attenuated recovery of focal adhesions and may increase the risk of in vivo breast cancer cell dissemination from primary lesions to distant sites.

3.3   Altered PRA:B Ratio Leads to Aberrant Mammary Gland Development in Mice

A role for progesterone in mammary gland tumorigenesis has been clearly demonstrated in studies using PR knockout mice (Lydon et al. 1999), and there is evidence from transgenic mice that changes in the balance of PRA:PRB leads to altered biology in tissues expressing both isoforms (Shyamala et al. 1998, 2000). Over-expression of PRA in mouse mammary tissues resulted in extensive mammary epithelial hyperplasia, increased ductal side-branching and disruption of the basement membrane, suggesting that a balanced expression of PRA and PRB is necessary for normal mammary gland development and function (Shyamala et al. 1998). In contrast, over-expression of PRB in the same model system leads to inhibition of ductal elongation and decreased lobular alveolar development (Shyamala et al. 2000).

4   Nuclear Location and Gene Expression

The nuclear architecture of a cell specifies subnuclear location and topological control of gene expression. The compartmentalised structure of the nucleus enforces specific spatial restraints within which the various functions in the nucleus occur, and correct assembly of components within these nuclear microenvironments is required for appropriate function. Discrete subnuclear locations have been described for a number of nuclear receptors, such as ER-α (Htun et al. 1999), and the androgen (Tomura et al. 2001; Kawate et al. 2005), mineralocorticoid (Nishi et al. 2001) and glucocorticoid receptors (McNally et al. 2000), and various steroid receptor-associated proteins have been shown to localise into discrete domains. The subnuclear position of a transcription factor, its associated co-factors, post-translational modifications, and the location, assembly and composition of the transcriptional apparatus all direct the fidelity and specificity of the transcriptional outcome. Nuclear structure is frequently disrupted in tumours, and nuclear features of cancer are useful clinical indicators.

4.1 Focal PRA and PRB Co-locate in Nuclear Foci in Normal Tissues

Nuclear receptors are reported to move into nuclear aggregates or foci in the presence of ligand (Racz and Barsony 1999; Prufer et al. 2000), and our recent studies in human breast and endometrial tissue also showed PR capable of focus formation. We observed PRA and PRB to be either distributed evenly throughout the nucleus in a diffuse, fine granular pattern, or to form discrete nuclear foci (Arnett-Mansfield et al. 2004). PR distribution into foci corresponded with the luteal phase of the menstrual cycle, where serum progesterone levels were high and PR activity was at its maximum. Conversely there was an even distribution of PRA and PRB during the follicular phase of the menstrual cycle in the absence of progesterone. There was a significant, inverse relationship between even and focal distribution, suggesting that PRA and PRB moved from an even to a focal distribution under the influence of luteal phase hormones (Fig. 6). These observations were further supported in mice where, during the progesterone peak of proestrus, PR localised into nuclear foci and, in ovariectomised mice, PR formed foci after exposure to oestrogen plus progesterone, but not under the influence of oestrogen alone (Arnett-Mansfield et al. 2007).

4.1.1 Ligand Binding Is Required for Nuclear Redistribution of PR

To directly test the role of ligand in PR foci formation, we have investigated newly expressed PR proteins in T-47D breast cancer cells for the presence of nuclear foci by taking advantage of the ability of the natural ligand progesterone to down-regulate PR, followed by the subsequent re-expression of nascent PR protein upon medium replacement (Arnett-Mansfield et al. 2007). The subnuclear distribution of newly synthesised PR, in an environment deprived of hormones, was predominantly even with very little or no focally distributed PRA or PRB, suggesting that in the absence of ligand, the receptor is distributed diffusely throughout the nucleus. When cells expressing nascent PR were treated with the synthetic progestin ORG2058, PRA and PRB redistributed into numerous, distinct nuclear foci. PRA and PRB were observed both in separate

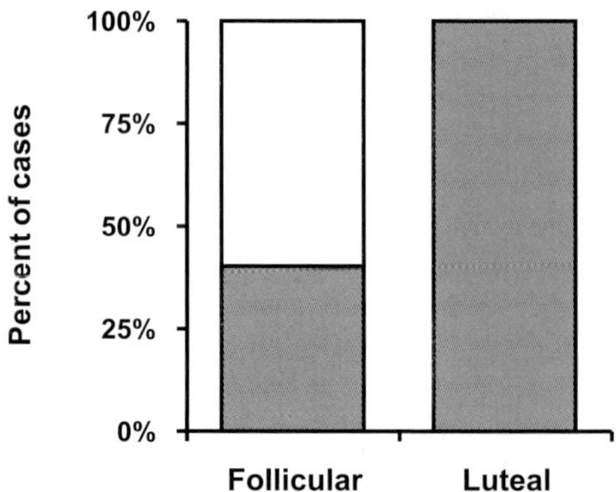

Fig. 6. Subnuclear distribution of PR in the normal breast. PR distribution was examined in a small cohort of normal human breast tissues, for which menstrual cycle stage was determined from matched endometrial specimens. PR distribution was scored as focal (*shaded*) or even (*open*). The proportion of cases falling into each category in the follicular (*n*=5) and luteal (*n*=3) phases of the menstrual cycle are shown

foci and co-located within the same focus, in accordance with our previous observations in mouse and human tissues (Arnett-Mansfield et al. 2004, 2007). This supported the notion that there was progestin-induced movement of PR into foci in T-47D cells. Similar results were seen in PR-positive MCF-7 breast cancer cells and ECC-1 endometrial cancer cells. Taken together, the distribution of PR into nuclear foci in cancer cell lines appeared to be an effect of ligand binding.

4.1.2　Nuclear Foci Contain PR Dimers

The fact that PR foci formation depends on the presence of ligand suggested that foci may represent activated PRA and PRB heterodimers and/or homodimers. To investigate this hypothesis further we constructed cyan and yellow fluorescent protein (CFP and YFP)-tagged PRA and PRB fusion proteins and transiently expressed them in PR-negative U-2 OS osteosarcoma cells (Arnett-Mansfield et al. 2007). The interaction of the two proteins was determined by estimating fluorescence resonance energy transfer (FRET).

The transfected cells were progestin-treated (ORG2058) or treated with vehicle, and expression and interaction between CFP and YFP fusions imaged using FRET. After correction for background and CFP and YFP crosstalk, a colour-encoded image was created that represented the level of FRET signal for each point within the cell. In vehicle-treated cells the fluorescent PR fusion proteins were evenly distributed in the nucleus and foci were rarely observed. Exposure to ORG2058 gave rise to a notable redistribution of the fluorescently labelled PR proteins into prominent foci, and foci were detectable when either one or both PR isoforms were present. A significant increase in FRET signal was detected in the nuclei of cells treated with ORG2058 when compared to those of vehicle-treated controls. In cells expressing both PRA and PRB the FRET signal in ORG2058-treated whole nuclei was increased two- to threefold compared to the signal from control nuclei, demonstrating that the regions within progestin-treated nuclei, which contained PR foci, were also the regions containing the highest density of FRET-producing PR dimers. Moreover, the foci in these cells had a FRET signal three- to four-fold higher than vehicle control.

In summary, FRET measurements using transfected fluorescent PR fusion proteins confirmed that PR foci were highly ligand-dependent and represent dimerised PR, and that the highest levels of FRET in progestin-treated cells were measured in foci for both PRA or PRB homodimers and for PRA-PRB heterodimers. The highest concentration of PRA and PRB dimers was found upon ligand binding and in nuclear foci.

4.1.3 Inhibition of Co-activator Recruitment and Transcription Prevent PR Movement into Foci

FRET analysis (Arnett-Mansfield et al. 2007) has demonstrated that PR foci contain the highest concentration of PR dimers and it is likely, therefore, that focally distributed PR isoforms are involved with transcription. To test this theory we treated T-47D breast cancer cells with ligand to increase PR foci, and examined the effects of transcriptional inhibitors on the focus forming ability of PR (Arnett-Mansfield et al. 2007). Roscovitine, a cyclin-dependent kinase inhibitor, has been shown to inhibit PR-dependent recruitment of the co-activator SRC-1. Exposure of T-47D cells to roscovitine alone resulted in a granular distribution of PR that was evenly distributed across the nucleus. Co-treatment of cells with roscovitine and ORG2058, however, completely abolished PR foci, demonstrating that the formation of ligand-dependent PR-containing transcriptional complexes was correlated with the presence of PR foci.

To further determine whether inhibition of transcription inhibited movement of PR into foci, T-47D cells were treated with ORG2058 in the presence or absence of the transcriptional inhibitor actinomycin D, and PR isoform distribution was detected by immunofluorescence. As previously observed, PR distribution in vehicle-treated cells was predominantly even and treatment with ORG2058 resulted in a highly focal pattern of PR detection. However, in the presence of actinomycin D, PR foci were abolished and an even nuclear distribution of PR was observed in both vehicle and ORG2058-treated cells. These results showed that either inhibiting recruitment of a transcriptional co-activator (SRC-1) or inhibiting transcription directly (actinomycin D) blocked formation of PR foci, suggesting that movement of PR into foci is associated with transcriptional regulation.

4.1.4 PR Foci Are Associated with Active Transcription

Incorporation of BrUTP into nascent RNA transcripts and subsequent dual detection with PR confirmed that PR foci are associated with active transcription (Arnett-Mansfield et al. 2007). Nascent transcripts were detected by incorporating BrUTP into run-on transcripts in T-47D breast cancer cells treated with ORG2058 or vehicle. Cells dual stained for PR and BrdU showed a distinct pattern of overlap in PR foci and nascent transcripts only in ORG2058-treated cells, initially detected after 5 min BrUTP incorporation, and became more prominent after 30 min incorporation.

To further support the notion that PR foci are associated with active transcription, additional experiments using dual immunofluorescent techniques to detect PR and the basic transcription factor RNA polymerase II were carried out. They demonstrated that co-localisation of PRA or PRB and RNAPolII was observed only in the presence of ligand (Arnett-Mansfield et al. 2007). Moreover, in progestin-treated T-47D cells PR foci co-localised with the transcriptional co-activator p300, which is known to associate with ligand-activated PR on chromatin, providing additional evidence that PR foci represent sites of transcriptional activity.

4.2   PR Foci in Breast Cancers

PR foci were observed in human PR-positive breast tumours; only around 50% of cases, however, displayed focal distribution despite the fact that all expressed PR (Arnett-Mansfield et al. 2007). The level of PR in breast cancers was a poor indicator of the ability to form foci. Focal formation was most prevalent in lesions with a predominance of PRA, or in tumours where both isoforms were equally expressed. It was less frequent in the relatively small proportion of tumours in which PRB was over-expressed.

4.2.1   PR Foci in Breast Cancer: Influence of Menopausal Status

In the normal endometrium of both human and mouse, PR foci were most prevalent during periods of high circulating progesterone levels (Arnett-Mansfield et al. 2004). PR foci in breast cancers, however, were equally likely to be found in pre- and post-menopausal women, suggesting that they were not influenced by the systemic endocrine environment and circulating ligand levels. This is in marked contrast to normal breast tissues in which, as observed in the endometrium, focal formation was most evident during the progesterone-dominated luteal phase of the menstrual cycle (Arnett-Mansfield et al. 2004).

4.2.2   The Requirement for Ligand in PR Foci Formation Is Lost in Cancers

In the cancers of hormone-responsive tissues, PR foci appeared to be ligand-independent, as they could be observed, as mentioned above, under physiological conditions associated with low circulating hormones and limited availability of progesterone. To explore this further we have examined in situ normal endometrial tissue found adjacent to tumour (Arnett-Mansfield et al. 2004). The endometrial cancer cells were observed to have distinct focal PR expression, whilst in normal adjacent tissue PR was distributed evenly throughout the nucleus. This clearly demonstrated that low levels of circulating progesterone that in normal cells were insufficiently high to drive PR into foci were able to do so in malignant cells, and substantiated the view that the need for ligand to drive PR into foci was lost in cancers.

4.2.3 PR Foci in Cancers Are Larger

There is evidence in endometrial cancer that PR foci are larger than the foci found in normal tissue (Arnett-Mansfield et al. 2007). Using confocal microscopy, the median length of cancer-associated PR foci was shown to be significantly longer compared to foci in normal cells. The larger size of PR foci in cancers suggested that PR may associate in foci with a more numerous and/or different accompaniment of protein partners. It may also be related to alterations in chromatin structure, a well-established feature of malignancy.

4.2.4 Chromatin Structure Involved in PR Foci Formation

PR foci localised to regions of active chromatin in the cell, suggesting that they corresponded to transcriptionally active PR (Arnett-Mansfield et al. 2007). Moreover, there is in vitro evidence that formation of PR foci is closely related to maintenance of the chromatin structure. Treatment of T-47D cells with a histone deacetylase inhibitor (TSA, trichostatin A) that remodels chromatin revealed an association with two distinct changes in PR foci formation. First, PR foci were increased in number with TSA treatment alone, suggesting a relaxation of ligand dependence needed for PR to move into foci in cells with disrupted chromatin. Second, when TSA-treated cells were exposed to progestin, the foci formed were significantly larger than those in non-TSA-treated cells. This implied that maintenance of the size of PR foci was controlled by the physical association of PR at specific chromatin locations. It is known that chromatin remodelling is important in PR activation of target genes, and these data showed clearly that chromatin integrity plays an important role in normal PR foci formation.

5   Summary

Although breast cancer is usually diagnosed later in life, normal breast tissue is likely to undergo mutational changes early in a woman's life, during a period of cyclical exposure to ovarian hormones, thus providing an opportunity for progesterone to influence breast cancer development.

In normal breast tissue PRA and PRB are expressed at similar levels; an imbalance of PRA and PRB, however, occurs early in breast cancer development and is commonly seen in premalignant lesions. As the normal breast expresses similar relative levels of PRA and PRB it is likely to be the heterodimer species that is active in this tissue. However, in many breast cancers relative levels of PRA and PRB differ widely from those seen in normal cells, and a frequent predominance of one isoform suggests the homodimer to be the active PR dimer species in malignancy.

Progesterone acts through PR and the receptors move into nuclear foci in the presence of ligand. Gene expression analyses have shown that alterations in PRA and PRB expression give rise to progesterone regulation of new and aberrant target genes, and increasing progestin responsiveness in signalling pathways associated with cell shape, adhesion and survival. Moreover, the normal ligand-dependency of PR movement into foci is lost in breast cancer tissues, and an increase in focal size suggests that alterations in the assembly of co-modulators occur prior to PR activation of target genes.

In summary disruption of PRA and PRB expression is an early event in breast cancer that results in aberrant regulation of progestin-responsive target genes. Altered cell function consequent to aberrant progesterone-mediated gene expression is a potential mechanism for progesterone signalling in breast cancer development, and supports the existing evidence of an association between progesterins and breast cancer.

References

Anderson E (2002) The role of oestrogen and progesterone receptors in human mammary development and tumorigenesis. Breast Cancer Res 4:197–201

Arnett-Mansfield RL, DeFazio A, Mote PA, Clarke CL (2004) Subnuclear distribution of progesterone receptors A and B in normal and malignant endometrium. J Clin Endocrinol Metab 89:1429–1442

Arnett-Mansfield RL, Graham JD, Hanson AR, Mote PA, Gompel A, Scurr LL, Gava N, de Fazio A, Clarke CL (2007) Focal subnuclear distribution of progesterone receptor is ligand dependent and associated with transcriptional activity. Mol Endocrinol 21:14–29

Bain DL, Franden MA, McManaman JL, Takimoto GS, Horwitz KB (2001) The N-terminal region of human progesterone B-receptors: biophysical and biochemical comparison to A-receptors. J Biol Chem 276:23825–23831

Beral V, Million Women Study Collaborators (2003) Breast cancer and hormone-replacement therapy in the Million Women Study. Lancet 362:419–427

Bocchinfuso WP, Korach KS (1998) Mammary gland development and tumorigenesis in estrogen receptor knockout mice. J Mammary Gland Biol Neoplasia 2:323–334

Brisken C, Park S, Vass T, Lydon JP, O'Malley BW, Weinberg RA (1998) A paracrine role for the epithelial progesterone receptor in mammary gland development. Proc Natl Acad Sci USA 95:5076–5081

Brody JG, Rudel RA, Michels KB, Mousich KB, Bernstein L, Attfield KR, Gray S (2007) Environmental pollutants, diet, physical activity, body size, and breast cancer: where do we stand in research to identify opportunities for prevention? Cancer 109:2627–2634

Clavel-Chapelon F, Gerber M (2002) Reproductive factors and breast cancer risk. Do they differ according to age at diagnosis? Breast Cancer Res Treat 72:107–115

Conneely OM, Mulac-Jericevic B, Lydon JP, De Mayo FJ (2001) Reproductive functions of the progesterone receptor isoforms: lessons from knock-out mice. Mol Cell Endocrinol 179:97–103

Dowsett M, Cuzick J, Wale C, Howell T, Houghton J, Baum M (2005) Retrospective analysis of time to recurrence in the ATAC trial according to hormone receptor status: an hypothesis-generating study. J Clin Oncol 23:7512–7517

Garcia SB, Park HS, Novelli M, Wright NA (1999) Field cancerization, clonality, and epithelial stem cells: the spread of mutated clones in epithelial sheets. J Pathol 187:61–81

Germain P, Altucci L, Bourguet W, Rochette-Egly C, Gronemeyer H (2003) Nuclear receptor superfamily: principles of signalling. Pure Appl Chem 75:1619–1664

Giangrande PH, Kimbrel EA, Edwards DP, McDonnell DP (2000) The opposing transcriptional activities of the two isoforms of the human progesterone receptor are due to differential cofactor binding. Mol Cell Biol 20:3102–3115

Goepfert TM, McCarthy M, Kittrell FS, Stephens C, Ullrich RL, Brinkley BR, Medina D (2000) Progesterone facilitates chromosome instability (aneuploidy) in p53 null normal mammary epithelial cells. FASEB J 14:2221–2229

Goepfert TM, Adigun YE, Zhong L, Gay J, Medina D, Brinkley WR (2002) Centrosome amplification and overexpression of aurora A are early events in rat mammary carcinogenesis. Cancer Res 62:4115–4122

Gong G, DeVries S, Chew KL, Cha I, Ljung BM, Waldman FM (2001) Genetic changes in paired atypical and usual ductal hyperplasia of the breast by comparative genomic hybridization. Clin Cancer Res 7:2410–2414

Graham JD, Clarke CL (1997) Physiological action of progesterone in target tissues. Endocr Rev 18:502–519

Graham JD, Clarke CL (2002) Expression and transcriptional activity of progesterone receptor A and progesterone receptor B in mammalian cells. Breast Cancer Res 4:187–190

Graham JD, Yager ML, Hill HD, Byth K, O'Neill GM, Clarke CL (2005) Altered progesterone receptor isoform expression remodels progestin responsiveness of breast cancer cells. Mol Endocrinol 19:2713–2735

Graham JD, Yeates C, Balleine RL, Harvey SS, Milliken JS, Bilous AM, Clarke CL (1995) Characterisation of progesterone receptor A and B protein expression in human breast cancer. Cancer Res 55:5063–5068

Greendale GA, Reboussin BA, Slone S, Wasilauskas C, Pike MC, Ursin G (2003) Postmenopausal hormone therapy and change in mammographic density. J Natl Cancer Inst 95:30–37

Hopp TA, Weiss HL, Hilsenbeck SG, Cui Y, Allred DC, Horwitz KB, Fuqua SAW (2004) Breast cancer patients with progesterone receptor PR-A-rich tumors have poorer disease-free survival rates. Clin Cancer Res 10:2751–2760

Howard BA, Gusterson BA (2000) Human breast development. J Mammary Gland Biol Neoplasia 5:119–137

Htun H, Holth LT, Walker D, Davie JR, Hager GL (1999) Direct visualization of the human estrogen receptor alpha reveals a role for ligand in the nuclear distribution of the receptor. Mol Biol Cell 10:471–486

Huse B, Verca SB, Matthey P, Rusconi S (1998) Definition of a negative modulation domain in the human progesterone receptor. Mol Endocrinol 12:1334–1342

Isaksson E, Wang H, Sahlin L, von Schoultz B, Cline JM, von Schoultz E (2003) Effects of long-term HRT and tamoxifen on the expression of progesterone receptors A and B in breast tissue from surgically postmenopausal cynomolgus macaques. Breast Cancer Res Treat 79:233–239

Kastner P, Krust A, Turcotte B, Stropp U, Tora L, Gronemeyer H, Chambon P (1990) Two distinct estrogen-regulated promoters generate transcripts encoding the two functionally different human progesterone receptor forms A and B. EMBO J 9:1603–1614

Kawate H, Wu Y, Ohnaka K, Tao RH, Nakamura K, Okabe T, Yanase T, Nawata H, Takayanagi R (2005) Impaired nuclear translocation, nuclear matrix targeting, and intranuclear mobility of mutant androgen receptors carrying amino acid substitutions in the deoxyribonucleic acid-binding domain derived from androgen insensitivity syndrome patients. J Clin Endocrinol Metab 90:6162–6169

Larson PS, de las Morenas A, Bennett SR, Cupples LA, Rosenberg CL (2002) Loss of heterozygosity or allele imbalance in histologically normal breast epithelium is distinct from loss of heterozygosity or allele imbalance in co-existing carcinomas. Am J Pathol 161:283–290

Larson PS, de las Morenas A, Cupples LA, Huang K, Rosenberg CL (1998) Genetically abnormal clones in histologically normal breast tissue. Am J Pathol 152:1591–1598

Lee SA, Ross RK, Pike MC (2005) An overview of menopausal oestrogen-progestin hormone therapy and breast cancer risk. Br J Cancer 92:2049–2058

Longacre TA, Bartow SA (1986) A correlative morphologic study of human breast and endometrium in the menstrual cycle. Am J Surg Pathol 10:382–393

Lydon JP, Ge G, Kittrell FS, Medina D, O'Malley BW (1999) Murine mammary gland carcinogenesis is critically dependent on progesterone receptor function. Cancer Res 59:4276–4284

McDonnell DP, Goldman ME (1994) RU486 exerts antiestrogenic activities through a novel progesterone receptor A form-mediated mechanism. J Biol Chem 269:11945–11949

McGowan EM, Saad S, Bendall LJ, Bradstock KF, Clarke CL (2004) Effect of progesterone receptor A predominance on breast cancer cell migration into bone marrow fibroblasts. Breast Cancer Res Treat 83:211–220

McGowan EM, Weinberger RP, Graham JD, Hill HD, Hughes JA, O'Neill GM, Clarke CL (2003) Cytoskeletal responsiveness to progestins is dependent on progesterone receptor A levels. J Mol Endocrinol 31:241–253

McKenna NJ, O'Malley BW (2002) Combinatorial control of gene expression by nuclear receptors and coregulators. Cell 108:465–474

McNally JG, Muller WG, Walker D, Wolford R, Hager GL (2000) The glucocorticoid receptor: rapid exchange with regulatory sites in living cells. Science 287:1262–1265

McTiernan A, Martin CF, Peck JD, Aragaki AK, Chlebowski RT, Pisano ED, Wang CY, Brunner RL, Johnson KC, Manson JE, Lewis CE, Kotchen JM, Hulka BS, Women's Health Initiative Mammogram Density Study I (2005) Estrogen-plus-progestin use and mammographic density in postmenopausal women: women's health initiative randomized trial. J Natl Cancer Inst 97:1366–1376

Mote PA, Balleine RL, McGowan EM, Clarke CL (1999) Co-localization of progesterone receptors A and B by dual immunofluorescent histochemistry in human endometrium during the menstrual cycle. J Clin Endocrinol Metab 84:2963–2971

Mote PA, Balleine RL, McGowan EM, Clarke CL (2000) Heterogeneity of progesterone receptors A and B expression in human endometrial glands and stroma. Hum Reprod 15:48–56

Mote PA, Bartow S, Tran N, Clarke CL (2002) Loss of co-ordinate expression of progesterone receptors A and B is an early event in breast carcinogenesis. Breast Cancer Res Treat 72:163–172

Mote PA, Leary JA, Avery KA, Sandelin K, Chenevix-Trench G, Kirk JA, Clarke CL (2004) Germ-line mutations in BRCA1 or BRCA2 in the normal breast are associated with altered expression of estrogen-responsive proteins and the predominance of progesterone receptor A. Genes Chromosomes Cancer 39:236–248

Mulac-Jericevic B, Mullinax RA, DeMayo FJ, Lydon JP, Conneely OM (2000) Subgroup of reproductive functions of progesterone mediated by progesterone receptor-B isoform. Science 289:1751–1754

Mulac-Jericevic B, Lydon JP, DeMayo FJ, Conneely OM (2003) Defective mammary gland morphogenesis in mice lacking the progesterone receptor B isoform. Proc Natl Acad Sci USA 100:9744–9749

Nishi M, Ogawa H, Ito T, Matsuda KI, Kawata M (2001) Dynamic changes in subcellular localization of mineralocorticoid receptor in living cells: in comparison with glucocorticoid receptor using dual-color labeling with green fluorescent protein spectral variants. Mol Endocrinol 15:1077–1092

Osborne CK, Schiff R, Arpino G, Lee AS, Hilsenbeck VG (2005) Endocrine responsiveness: understanding how progesterone receptor can be used to select endocrine therapy. Breast 14:458–465

Park CC, Henshall-Powell RL, Erickson AC, Talhouk R, Parvin B, Bissell MJ, Barcellos-Hoff MH (2003) Ionizing radiation induces heritable disruption of epithelial cell interactions. Proc Natl Acad Sci USA 100:10728–10733

Prufer K, Racz A, Lin GC, Barsony J (2000) Dimerization with retinoid X receptors promotes nuclear localization and subnuclear targeting of vitamin D receptors. J Biol Chem 275:41114–41123

Racz A, Barsony J (1999) Hormone-dependent translocation of vitamin D receptors is linked to transactivation. J Biol Chem 274:19352–19360

Richer JK, Jacobsen BM, Manning NG, Abel MG, Wolf DM, Horwitz KB (2002) Differential gene regulation by the two progesterone receptor isoforms in human breast cancer cells. J Biol Chem 277:5209–5218

Rossouw J, Anderson GL, Prentice RL, LaCroix AZ, Kooperberg C, Stefanick ML, Jackson RD, Beresford SA, Howard BV, Johnson KC, Kotchen JM, Ockene J, Writing Group for the Women's Health Initiative Investigators (2002) Risks and benefits of estrogen plus progestin in healthy postmenopausal women: principal results From the Women's Health Initiative randomized controlled trial. JAMA 288:321–333

Santen RJ (2003) Risk of breast cancer with progestins: critical assessment of current data. Steroids 68:953–964

Santen RJ, Pinkerton J, McCartney C, Petroni GR (2001) Risk of breast cancer with progestins in combination with estrogen as hormone replacement therapy. J Clin Endocrinol Metab 86:16–23

Sartorius CA, Melville MY, Hovland AR, Tung L, Takimoto GS, Horwitz KB (1994) A third transactivation function (AF3) of human progesterone receptors located in the unique N-terminal segment of the B-isoform. Mol Endocrinol 8:1347–1360

Schneider W, Ramachandran C, Satyaswaroop PG, Shyamala G (1991) Murine progesterone receptor exists predominantly as the 83-kilodalton 'A' form. J Steroid Biochem Mol Biol 38:285–291

Shyamala G, Yang X, Cardiff RD, Dale E (2000) Impact of progesterone receptor on cell-fate decisions during mammary gland development. Proc Natl Acad Sci USA 97:3044–3049

Shyamala G, Yang X, Silberstein G, Barcellos-Hoff MH, Dale E (1998) Transgenic mice carrying an imbalance in the native ratio of A to B forms of progesterone receptor exhibit developmental abnormalities in mammary glands. Proc Natl Acad Sci USA 95:696–701

Smalley M, Ashworth A (2003) Stem cells and breast cancer: a field in transit. Nat Rev Cancer 3:832–844

Soderqvist G, Isaksson E, von Schoultz B, Carlstrom K, Tani E, Skoog L (1997) Proliferation of breast epithelial cells in healthy women during the menstrual cycle. Am J Obstet Gynecol 176:123–128

Sorlie T, Perou CM, Tibshirani R, Aas T, Geisler S, Johnsen H, Hastie T, Eisen MB, van de Rijn M, Jeffrey SS, Thorsen T, Quist H, Matese JC, Brown PO, Botstein D, Eystein Lonning P, Borresen-Dale AL (2001) Gene expression patterns of breast carcinomas distinguish tumor subclasses with clinical implications. Proc Natl Acad Sci USA 98:10869–10874

Tomura A, Goto K, Morinaga H, Nomura M, Okabe T, Yanase T, Takayanagi R, Nawata H (2001) The subnuclear three-dimensional image analysis of androgen receptor fused to green fluorescence protein. J Biol Chem 276:28395–28401

Topper YJ, Freeman CS (1980) Multiple hormone interactions in the developmental biology of the mammary gland. Physiol Rev 60:1049–1106

Tsai MJ, O'Malley BW (1994) Molecular mechanisms of action of steroid/thyroid receptor superfamily members. Annu Rev Biochem 63:451–486

Tung L, Mohamed MK, Hoeffler JP, Takimoto GS, Horwitz KB (1993) Antagonist-occupied human progesterone B-receptors activate transcription without binding to progesterone response elements and are dominantly inhibited by A-receptors. Mol Endocrinol 7:1256–1265

Vegeto E, Shahbaz MM, Wen DX, Goldman ME, O'Malley BW, McDonnell DP (1993) Human progesterone receptor A form is a cell- and promoter-specific repressor of human progesterone receptor B function. Mol Endocrinol 7:1244–1255

Venkitaraman AR (2002) Cancer susceptibility and the functions of BRCA1 and BRCA2. Cell 108:171–182

Wen DX, Xu YF, Mais DE, Goldman ME, McDonnell DP (1994) The A and B isoforms of the human progesterone receptor operate through distinct signaling pathways within target cells. Mol Cell Biol 14:8356–8364

Wiehle RD, Helftenbein G, Land H, Neumann K, Beato M (1990) Establishment of rat endometrial cell lines by retroviral mediated transfer of immortalizing and transforming oncogenes. Oncogene 5:787–794

Yue W, Wang JP, Li Y, Bocchinfuso WP, Korach KS, Devanesan PD, Rogan E, Cavalieri E, Santen RJ (2005) Tamoxifen versus aromatase inhibitors for breast cancer prevention. Clin Cancer Res 11:925s–930s

Ernst Schering Foundation Symposium Proceedings, Vol. 1, pp. 109–126
DOI 10.1007/2789_2007_058
© Springer-Verlag Berlin Heidelberg
Published Online: 17 December 2007

Inhibition of Mammary Tumorigenesis by Estrogen and Progesterone in Genetically Engineered Mice

D. Medina(✉), F.S. Kittrell, A. Tsimelzon, S.A.W. Fuqua

Department of Molecular and Cellular Biology, and Baylor Breast Center, Baylor College of Medicine, One Baylor Plaza, 77030 Houston, USA
email: *dmedina@bcm.tmc.edu*

Abstract. Estrogen and progesterone play a critical role in normal and neoplastic development of the mammary gland. A long duration of estrogen and progesterone exposure is associated with increased breast cancer risk, and a short duration of the same doses of these hormones is associated with a reduced breast cancer risk. The protective effects of estrogen and progesterone have been extensively studied in animal models. Several studies have demonstrated that these hormones induce persistent and long-lasting alterations in gene expression in the mammary epithelial cells. In the experiments discussed herein, the protective effect of estrogen and progesterone is shown to occur in genetically engineered mice (the p53-null mammary gland). The protective effect is associated with a decrease in cell proliferation. The effects of hormones seem to manifest as a delay in premalignant progression. In the nontumor-bearing glands of hormone treated mice, premalignant foci are present at the time the control glands are actively developing mammary tumors. If the hormone-treated cells are transplanted from the treated host to the untreated host, the cells resume their predetermined tumorigenic potential. The protective effect reflects both host-mediated factors (either stroma-determined or systemic factors) and mammary epithelial intrinsic changes. If normal, untreated p53 cells are transplanted into a host that has been previously treated with a short dose of hormones, the cells exhibit a significant delay in tumorigenesis. The relative contributions of host-mediated factors and mammary cell intrinsic factors remain to be determined. Current studies are moving this research area from the biological to the molecular realm and from the rodent models to human studies and offer the potential for directing prevention efforts at specific molecular targets.

1   Introduction

Breast cancer is a major cancer of women in the United States and Western Europe, with approximately 213,000 new cases expected to occur in the United States in 2006 (American Cancer Society 2006). The rate of increase of breast cancer incidence slowed in the 1990s; the incidence of in situ breast lesions, mainly ductal carcinoma in situ (DCIS), increased during the same time. The mortality rate was relatively constant through the last quarter of the twentieth century, before showing a significant decrease starting in the early 1990s. Early diagnosis and continuing new therapeutic approaches have managed to prevent the epidemic from causing a concomitant increase in death. Nevertheless, the death of 41,000 women due to invasive breast cancer remains a sobering fact and indicates the need to understand this disease in greater depth and to develop new interventions, both preventative and therapeutic.

Current understanding of the central role of hormones in the genesis of breast cancer is based on over 100 years of studies. Beatson (1896) developed the first therapeutic treatment with bilateral oophorectomy, thus removing the source of steroid reproductive hormones. In current medical practice, the presence of estrogen receptor (ER) is the most widely used predictive factor of breast cancer response to treatment and represents the rationale for the use of selective estrogen receptor modulators (SERMs) (i.e., tamoxifen and raloxifene) and aromatase inhibitors (Jordan et al. 1991; Osborne 1998; Brodie et al. 1999; Goss 2004) as adjuvant therapy against breast cancer recurrence. Importantly, recent trials have documented that hormone replacement therapy that includes both estrogens and progestins imparts a greater breast cancer risk than estrogens alone (Santen 2003).

Ironically, although early menarche and total years of hormone exposure are risk factors for increased incidence of breast cancer, early age of first pregnancy ($\leq$20 years of age) is a strong protective factor (one-half risk compared with nulliparous women). The protective factor is especially observed in postmenopausal women, the period of peak incidence. Parity-induced protection against breast cancer is principally dependent on the timing of the first full-term pregnancy but also is affected by total number of pregnancies (Ahlgren et al. 2004). The protective effect of early first pregnancy has been repeatedly demonstrated in numerous epidemiological studies and provides a physiologically operative model to achieve practical and affordable prevention of breast cancer in humans (Henderson et al. 1991). The logic and rationale for understanding the molecular basis for hormone-mediated prevention of breast cancer are based on the consistent observations in human epidemiological studies and the strong confirmatory experiments in rodent breast cancer models (Medina 2005).

In the classical chemical carcinogen-induced mammary tumor models of rats and mice, there are two different experimental models that demonstrate parity/hormone-induced protection. In the first model, half of the animals undergo hormonal stimulation, the mammary gland is allowed to involute completely, and then the carcinogen is administered to the hormone-treated and age-matched virgin animals. This model is termed the "pre-treatment model" and most of the experiments use this experimental protocol. In the second model, the animals are treated with a carcinogen, and then half are exposed to hormone treatment for a specified time period. This model is referred to as the "post-treatment" model. The distinction is important because the underlying mechanisms for protection are likely to be different between the two models.

Recent studies have focused on using genetically engineered mouse models of breast cancer as these models are based on genetic changes observed in major subsets of human breast cancer and mimic closely the biological, histological, and chromosomal events present in the human disease. In the studies reported herein, we have used the p53-null mammary epithelial transplant model to elucidate the biological and molecular events associated with estrogen–progesterone (E/P)-induced protection against mammary tumorigenesis.

2　　Results

2.1　　Hormone-Mediated Prevention in p53-Null Mammary Epithelium

Experiment 1 examined the tumorigenic response of the p53-null mammary epithelium exposed to estrogen and progesterone combination at 2–4 weeks after transplantation (Fig. 1). In the untreated mice, 16/66 transplants (24%) produced tumors by 45 weeks after transplantation.

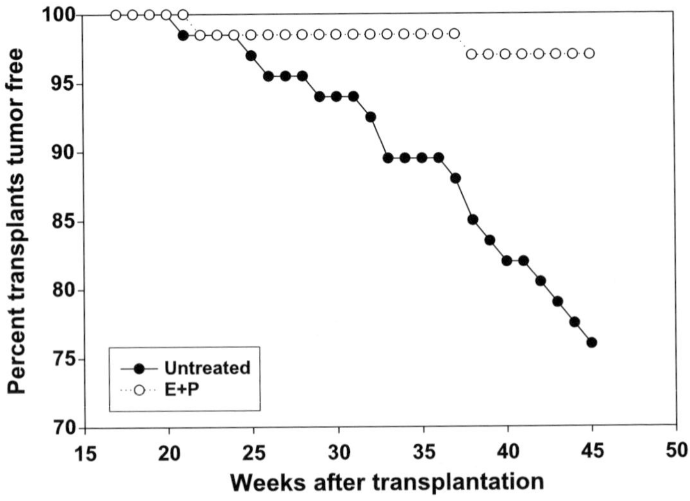

Fig. 1. The protective effect of short-term estrogen and progesterone treatment. Mice were treated with 50 µg estradiol and 20 mg progesterone at 5–7 weeks of age. The treated mice showed a significant reduction ($p<.05$) in mammary tumor incidence

In the hormone-treated mice, only 2/66 (3%) produced tumors ($p<.05$). This experiment conclusively demonstrated that a short-term hormone treatment can delay tumor development in a noncarcinogen model of mammary cancer.

Whole-mount analysis of the glands at 45 weeks after transplantation revealed extensive ductal hyperplasias in both control and hormone-treated glands. In the E/P-treated group, 13/31 glands (42%) contained hyperplastic lesions and in the untreated control group, 6/13 glands (46%) contained hyperplastic lesions. These data suggested that hormones act to block premalignant progression and not the onset of hyperplastic growth.

To directly test the malignant potential of the premalignant lesions, we transplanted the existing lesions into the cleared glands of 3-week-old mice. As shown in Fig. 2, the tumorigenic potentials of the two

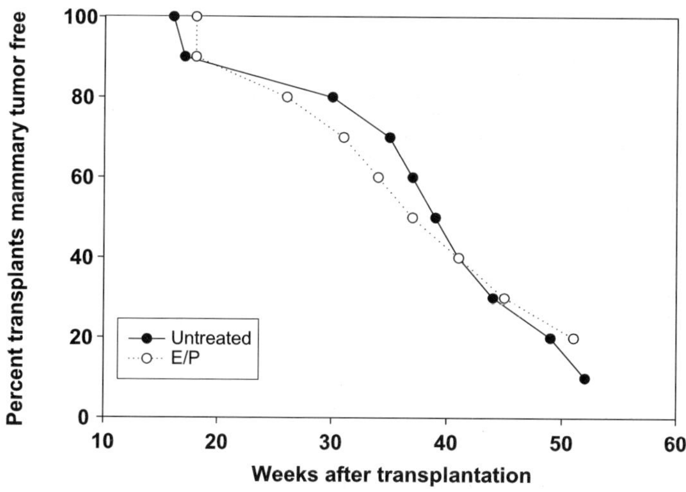

Fig. 2. The tumorigenic potential of hormone treated and untreated p53-null mammary epithelial cells. Samples of mammary tissue were taken from 45-week-old mice, both untreated and E/P-treated mice, and transplanted into the cleared fat pads of 3-week-old wildtype Balb/c mice. The mice were palpated weekly for tumors over a 52-week period. The tumor incidence was the same for both donor groups

donor groups were virtually identical. There were 46 transplants from the untreated control donor mice and 66 transplants from the E/P-treated donor mice.

2.2　Systemic Effect of Hormones

Experiment 2 tested the indirect effect of hormone pretreatment on tumorigenesis in p53-null mammary epithelium. In this experiment, mice were treated with E/P at age 5–7 weeks. The p53-null mammary epithelial cells were transplanted at either 3 or 11 weeks of age, thus comparing the tumorigenic potential of cells directly exposed vs indirectly ex-

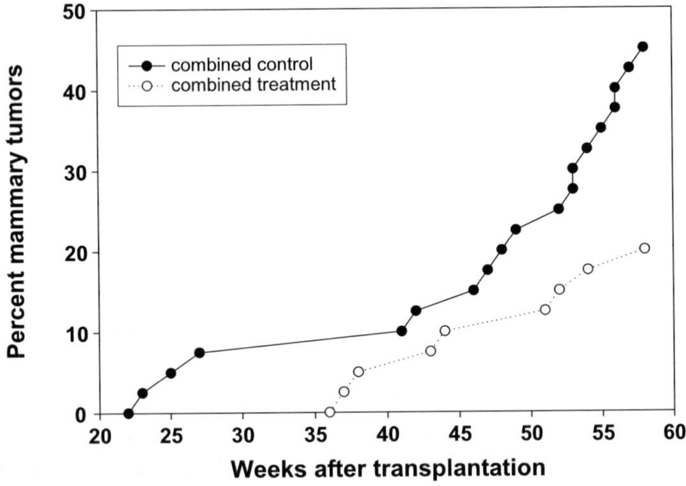

Fig. 3. The effect of hormonal treatment of the host on tumorigenesis in p53-null mammary epithelial transplants. Four groups of mice were cleared at 3 weeks of age and the treated groups given estrogen and progesterone at 5–7 weeks. Two groups of mice (one treated and one untreated) were implanted with mammary duct fragments at 3 weeks of age; the other two (one treated and one untreated) were implanted at 11 weeks, which was 4 weeks after removal of the hormones. Mammary transplants in both of the hormone-exposed hosts had a significant reduction in tumor incidence compared to untreated hosts. The graph shows combined treatment groups vs combined untreated groups

posed to the hormone treatment. The tumorigenic capability of the p53 mammary epithelial cells either directly exposed to E/P or exposed just to the E/P-treated host is shown in Fig. 3. The tumorigenic capability was the same whether the target epithelial cells were directly exposed to E/P or transplanted into an E/P-treated host environment (3/15 vs 4/20, respectively) ($p > .05$). The tumorigenic response was decreased from 45% (18/40) in the combined control groups to 20% (7/35) for the combined E/P treatment groups ($p < .05$) (Fig. 3).

The proliferation state of the mammary epithelial cells was assessed at 4–12 weeks after removal of the hormones in both experiments (Fig. 4). In experiment 1, at 4 and 8 weeks after hormone removal, the control transplants showed a BrdU-labeling index of 9.8% and 8.2%,

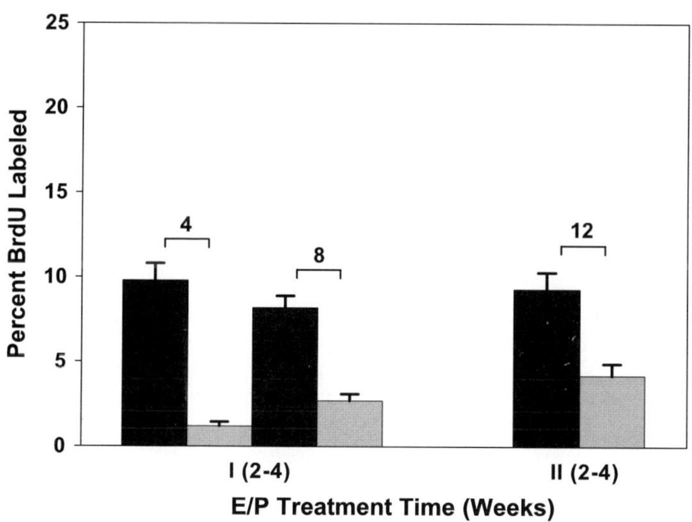

Fig. 4. The proliferation index (BrdU-labeling index) in hormone-treated p53-null mammary transplants. Transplants from both experiments (1 and 2) were assayed for BrdU retention (number labeled/500 cells). The *black bars* are untreated controls, the *clear bars* are hormone-treated. In experiment 2, the 12 weeks represents 8 weeks after transplantation. There were four transplants per group. All comparisons between treated and untreated groups were significantly different ($p < .05$)

                                                    D. Medina et al.

Table 1 Differentially expressed genes in hormone-treated p53-null mammary epithelial cells

Number	Gene symbol	Gene name	Fold change (Cont/EP)
1	Slc2a3	Solute carrier family 2 (facilitated glucose transporter), member 3	2.6
2	Per2	Period homolog 2 (Drosophila)	2.4
3	Epgn	Epithelial mitogen	2.3
4	Rasd1	RAS, dexamethasone-induced 1	2.1
5	Mgll	Monoglyceride lipase	2.1
6	Rrm2	Ribonucleotide reductase M2	2.1
7	Klf4	Kruppel-like factor 4	2.1
8	Mad2l1	MAD2 (mitotic arrest deficient, homolog)-like 1 (yeast)	2.0
9	Fosb	FBJ osteosarcoma oncogene B	1.9
10	Usp2	Ubiquitin specific protease 2	1.8
11	Rbp7	Retinol binding protein 7, cellular	1.7
12	Tef	Thyrotroph embryonic factor	1.7
13	Cdc42ep4	CDC42 effector protein (Rho GTPase binding) 4	1.6
14	Mrvldc1	MARVEL (membrane-associating) domain containing 1	1.6
15	Angptl2	Angiopoietin-like 2	1.6
16	Sh3bp5	SH3-domain binding protein 5 (BTK-associated)	1.5
17	Bace1	β-Site APP cleaving enzyme 1	1.5
18	Pdcd4	Programmed cell death 4	1.5
19	Fabp4	Fatty acid binding protein 4, adipocyte	1.5
20	Alas1	Aminolevulinic acid synthase 1	1.5
21	Nab2	Ngfi-A binding protein 2	1.5
22	Fmod	Fibromodulin	1.5
23	Twist2	Twist homolog 2 (Drosophila)	−1.4
24	Trib3	Tribbles homolog 3 (Drosophila)	−1.4
25	Klf7	Kruppel-like factor 7 (ubiquitous)	−1.4
26	Fusip1	FUS interacting protein (serine-arginine rich) 1	−1.4
27	Gpr109b	G protein-coupled receptor 109B	−1.4

Table 1 (continued)

Number	Gene symbol	Gene name	Fold change (Cont/EP)
28	Lrrfip1	Leucine rich repeat (in FLII) interacting protein 1	−1.4
29	Jub	Ajuba	−1.5
30	Tnip1	TNFAIP3 interacting protein 1	−1.5
31	Rdh10	Retinol dehydrogenase 10 (all-trans)	−1.5
32	Dsg2	Desmoglein 2	−1.5
33	Gars	Glycyl-tRNA synthetase	−1.5
34	Cpne1	Copine I	−1.5
35	Nars	Asparaginyl-tRNA synthetase	−1.6
36	Tars	Threonyl-tRNA synthetase	−1.6
37	Ptpns1	Protein tyrosine phosphatase, nonreceptor type substrate 1	−1.6
38	Iars	Isoleucine-tRNA synthetase	−1.6
39	Ctps	Cytidine 5′-triphosphate synthase	−1.7
40	Il18r1	Interleukin 18 receptor 1	−1.7
41	Tyki	Thymidylate kinase family LPS-inducible member	−1.7
42	Il1rn	Interleukin 1 receptor antagonist	−1.7
43	Cxcl10	Chemokine (C-X-C motif) ligand 10	−1.7
44	Mmp3	Matrix metalloproteinase 3	−1.8
45	Nupr1	Nuclear protein 1	−1.9
46	Tceb3bp1	Transcription elongation factor B polypeptide 3 binding protein 1	−1.9
47	Tacstd2	Tumor-associated calcium signal transducer 2	−1.9
48	Il1rl1	Interleukin 1 receptor-like 1	−1.9
49	Tnfrsf21	Tumor necrosis factor receptor superfamily, member 21	−1.9
50	Atp1b1	ATPase, Na^+/K^+ transporting, $\beta1$ polypeptide	−2.0
51	Gltp	Glycolipid transfer protein	−2.0
52	Asns	Asparagine synthetase	−2.0
53	Mmp13	Matrix metalloproteinase 13	−2.1
54	Ptgs2	Prostaglandin-endoperoxide synthase 2	−2.1

Table 1 (continued)

Number	Gene symbol	Gene name	Fold change (Cont/EP)
55	Snx9	Sorting nexin 9	−2.2
56	Rbm6	RNA binding motif protein 6	−2.2
57	Lss	Lanosterol synthase	−2.3
58	Lor	Loricrin	−2.3
59	Ifit1	Interferon-induced protein with tetratricopeptide repeats 1	−2.3
60	Dst	Dystonin	−2.4
61	Rgl1	Ral guanine nucleotide dissociation stimulator-like 1	−2.9
62	Mthfd2	Methylenetetrahydrofolate dehydrogenase (NAD+dependent), methenyltetrahydrofolate cyclohydrolase	−3.0
63	Cxcl5	Chemokine (C-X-C motif) ligand 5	−3.3

Cont/EP, control/estrogen-progesterone

respectively, compared to 1.5% and 1.8%, respectively, in the hormone-treated transplants ($p<.05$). In experiment 2, at 12 weeks after hormone removal (i.e., 8 weeks after transplantation into 11-week-old mice), the control transplants has a BrdU-labeling index of 9.5% vs 4.5% in the transplants in the hormone-treated mice ($p<.05$).

2.3 Gene Expression

From the samples in experiment 1, we analyzed the gene expression signature at 45 weeks after hormone exposure using Affymetrix microarray platform. This was a time when mammary tumors appeared in 24% of the controls and only 3% of the E/P-treated mice. Table 1 lists some of the more prominent genes that were up- or downregulated with a $p<.001$. There were 50 genes upregulated and 33 genes downregulated in the E/P-treated mammary cells compared to the untreated control cells. Of interest were genes involved in proliferation and membrane cytoskeletal function. Genes that were elevated in control cells include the growth factors *epigen* and *fos*, the membrane/cytoskeletal factor *fibromodulin*, and the transcription factor *Kruppel-like-4*. In contrast, genes that were elevated in E/P-treated cells include the tumor suppressor *nuclear protein 1* (*com-1*, *p8*), the membrane/cytoskeletal factors *desmoglein 2* and *matrix metalloproteinase 3* (*MMP3*) (*stromelysin*), the transcription factors *twist2* and *trib2*, and the cytokines *cxcl5* and *-10*. There were two general conclusions that were evident from this preliminary analysis. First, there were several genes that correlate with a reduced growth potential of the E/P-treated cells; namely, the increase in the tumor suppressor *com-1(p8)*, and the decrease in the growth factor *epigen* and *MMP3*. Second, none of the genes differentially expressed in the p53-null cells (although there were three closely related genes) are found in the extensively annotated pregnancy signature reported by Blakely et al. (2006, 2007).

3   Discussion

The experiments discussed herein provide three results important for the study and understanding of hormone-induced protection against breast cancer. First, they address the question of whether short-term hormone treatment can delay tumorigenesis in genetically engineered models of mammary cancer. All previous experiments, with the one exception, where radiation was used were performed in chemical carcinogen-treated rodent models. The results clearly demonstrate that a short-term treatment of estrogen and progesterone significantly delays tumorigenesis in the p53-null mammary epithelium. The p53-null epithelium represents a model where a major tumor suppressor gene is deleted and the biological and genetic properties of the premalignant and malignant stages mimic many features found in human breast cancer (Medina et al. 2002). Recently, hormone-induced protection has been also shown in another genetically engineered mouse model, the activated-neu model, which represents a model where overexpression of an oncogene induces tumors very rapidly and with high multiplicity (Rajkumar et al. 2007).

Second, the experiments address the question of whether hormone-induced refractoriness is targeted to the initiation stage or promotion stage of carcinogenesis. A few experiments have provided data on this issue, although without definitive conclusions (Nandi et al. 2005; Reddy et al. 2002; Yang et al. 1999). Previous results demonstrated that the hormone-induced refractory gland, upon autopsy, contains small and nonpalpable microtumors in the absence of palpable tumors. The frequency of the microtumors was less than or equal to that found in methylnitrosourea (MNU)-treated age-matched virgins (AMV). Additionally, the provocative experiments by Thordarson et al. (2001)—demonstrating that continuing the same doses of hormones for an extended period of time (20 rather than 3 weeks) results in abundant tumor development equivalent to the AMV—indicate that hormones can promote progression if administered continuously. A similar conclusion was reached in the parous mouse using pituitary isografts as the source of hormones (Swanson et al. 1995). The mouse experiments we summarize above reproduced the previous results in rats by demonstrating that morphologically premalignant lesions were present in the glands of the E/P-treated and control mice at the same frequency at 45 weeks of age. More importantly, the experiments demonstrated that the growth and tumorigenic potentials of these hyperplasias were identical when transplanted into the mammary glands of untreated control mice. The accumulated results in rat and mouse models indicate that there is a point in premalignant progression where the cells are no longer susceptible to the preventative effects of this hormone combination.

Finally, the experiments demonstrated conclusively that the effect of the hormones can be mediated in part by changes induced at the systemic level, in the mammary stroma, or both. This result was not entirely unexpected. This idea was presented by Nandi and his coworkers in studies using the carcinogen-induced rat mammary system (Thordarson et al. 1995). Attempts to test this hypothesis were only partially successful because the mammary tumor incidence of carcinogen-treated gland transplanted into parous vs nulliparous rats was very low for both donor groups, although there was a significant difference in the number of hyperplasias (Abrams et al. 1998). The results presented in the previous sections used a mouse model and demonstrated conclusively that the target mammary epithelial cells were responsive to hormone-induced changes at the systemic level, in mammary stroma, or both, as the target cells were never directly exposed to the elevated levels of estrogen and progesterone. Such an effect is not without precedent in the literature. Barcellos-Hoff and Ravani demonstrated that irradiated stroma can alter premalignant progression of a mouse mammary outgrowth line, COMMA-D (Barcellos-Hoff and Ravani 2000). Maffini demonstrated that chemical carcinogen-treated (MNU) stroma can enhance progression to mammary tumors of rat mammary cells (Maffini et al. 2004). Schedin et al. (2004) demonstrated that the extracellular matrix of mammary stroma from parous stroma delayed glandular morphogenesis.

The cellular and molecular mechanisms underlying the changes systemically or in the mammary stroma remain to be identified. Both altered transforming growth factor (TGF)-β signaling (Barcellos-Hoff and Ravani 2000) and growth hormone (GH) signaling (Thordarson et al. 1995) have been implicated. Circulating levels of prolactin appeared lower (not statistically significant) in the parous rat compared to the AMV (Thordarson et al. 1995). Growth hormone concentration was significantly reduced in parous rats as compared with AMV rats (Thordarson et al. 1995). The parous animals also had decreased levels of ER and epidermal growth factor receptor (EGF-R) (Thordarson et al. 1995). The reduction in the circulating levels of GH caused a reduced susceptibility of the parous rats to mammary carcinogenesis, possibly by decreasing the levels of ER, EGF-R, insulin-like growth factor (IGF)-1, or a combination of them in the mammary gland. Interestingly, we could not demonstrate altered systemic IGF-1 levels in our hormone-treated mice at 4 weeks after hormone removal (D. Medina and A. Lee, unpublished data). The role of the GH–IGF1 axis in hormone-induced protection is supported by two additional results (Thordarson et al. 2004a,b). First, MNU-induced mammary tumorigenesis is almost totally absent in GH-deficient spontaneous dwarf rats (SDR). SDR rats given GH or IGF-1 before and after MNU treatment exhibited a normal (GH) or near normal (IGF-1) tumorigenic response. Estradiol (E2) and progesterone (P4) treatment did not restore tumorigenesis. Furthermore, E2+P4 blocked the response to GH. SDR rats have a normal circulating concentration of prolactin (PRL), indicating PRL was not a factor in the low tumorigenic incidence (Thordarson et al. 2004a). Second, IGF-1 treatment of parous rats for 60 days after NMU-treatment increased the mammary tumor incidence compared to untreated parous rats and comparable to AMV (Thordarson et al. 2004b).

Several different groups have published extensive gene expression profiles of parous vs aged-matched virgin mammary epithelial cells in rodent models. The comprehensive studies by Chodosh and colleagues (D'Cruz et al. 2002; Blakely et al. 2006, 2007), used DNA microarray technology to identify differentially expressed genes common in the parous involuted glands in multiple strains of both rats and mice. The authors identified a gene signature of 47 differentially expressed genes common to both species that defined the parous involuted gland. The genes included increased expression of TGF-β3 and several of its downstream targets. The net result of this gene upregulation would be an inhibition for cell growth. Additionally, there was decreased expression of several growth factors that included IGF-1 and amphiregulin. The result of the decreased expression of these genes would be a lowered stimulus for proliferation. The net result is a low, steady state for growth. To our surprise, the 83 genes that were differentially expressed in the E/P-treated cells did not overlap with the genes established in the Chodosh signature for the parous gland. There were only three genes that were of similar families; *MMP3* and 13 and *Kruppel-like factor 7*, all upregulated in the E/P-treated cells. However, of the genes that were differentially expressed, the end result would be predicted to be a reduced proliferation potential, a phenotype directly demonstrated by the BrdU experiments. A reduced proliferation potential has been reported in the parous involuted glands of both rats and mice treated with chemical carcinogens (Sivaraman et al. 1998, 2001).

Why our microarray data are so different from the data of Chodosh and colleagues is difficult to understand at this time. The Balb/c mouse strain we used in our experiments is included in the mouse strains that were reported in Blakely et al. (2006). One major difference is that we used just E/P while the Blakely et al. studies used a full-term pregnancy. If we believe that the E/P treatment mimics pregnancy, then one would expect some degree of overlap in the gene signature. Finally, we are utilizing a very specific genetically engineered mouse model and it may be that the absence of p53 is preferentially determining this gene expression signature.

4 Conclusions

In summary, these studies provide further rationale for considering the use of short-term hormone exposure as a preventative modality, particularly in high-risk individuals. Despite the extensive documentation of the preventative potential of early full-term pregnancy and its mimicry by estrogen and progesterone, there is great resistance to the use of these hormones as a preventative modality. In part, the resistance is due to the overwhelming data that show prolonged exposure to estradiol and progesterone increases breast cancer risk (Chlebowski et al. 2003). This resistance might be mitigated by the recent data that indicate hormone replacement therapy with estrogen alone does not increase the risk for breast cancer (Stefanick et al. 2006). An understanding of the mechanisms underlying the preventative effects of specific hormone combinations and duration of exposure will provide a basis for targeted intervention suing this paradigm.

Acknowledgements. The original results discussed here were supported by NCI P01 CA064255. We greatly appreciate the technical assistance of Frances Kittrell, Anne Shepard, David Edwards, and Irma Parra, and the secretarial skills of Kathy Key.

References

Abrams TJ, Guzman RC, Swanson SM, Thordarson G, Talamantes F, Nandi S (1998) Changes in the parous rat mammary gland environment are involved in parity-associated protection against mammary carcinogenesis. Anticancer Res 18:4115–4121

Ahlgren M, Melbye M, Wohlfahrt J, Sorensen TL (2004) Growth patterns and the risk of breast cancer in women. N Engl J Med 351:1619–1626

American Cancer Society (2006) Cancer facts and figures 2006. http://www.cancer.org/downloads/STT/CAFF2006PWSecured.pdf. Accessed 10 Aug 2007

Barcellos-Hoff MH, Ravani SA (2000) Irradiated mammary gland stroma promotes the expression of tumorigenic potential by unirradiated epithelial cells. Cancer Res 60:1254–1260

Beatson GT (1896) On the treatment of inoperable cases of carcinoma of the mamma. Suggestions for a new method of treatment with illustrative cases. Lancet 2:104–107

Blakely CM, Stoddard AJ, Belka GK, Dugan KD, Notarfrancesco KL, Moody SE, D'Cruz CM, Chodosh LA (2006) Hormone-induced protection against mammary tumorigenesis is conserved in multiple rat strains and identifies a core gene expression signature induced by pregnancy. Cancer Res 66:6421–6431

Blakely CM, Stoddard AJ, Belka GK, Dugan KD, Notarfrancesco KL, Moody SE, D'Cruz CM, Chodosh LA (2007) Correction: pregnancy-induced protection against mammary tumorigenesis. Cancer Res 67:844–845

Brodie A, Lu Q, Liu Y, Long B (1999) Aromatase inhibitors and their antitumor effects in model systems. Endocr Relat Cancer 6:205–210

Chlebowski RT, Hendrix SL, Langer RD, Stefanick ML, Gass M, Lane D, Rodabough RJ, Gilligan MA, Cyr MG, Thomson CA, Khandekar J, Petrovitch H, McTiernan A (2003) Influence of estrogen plus progestin on breast cancer and mammography in healthy postmenopausal women: the Women's Health Initiative Randomized Trial. JAMA 289:3243–3253

D'Cruz CM, Moody SE, Master SR, Hartman JL, Keiper EA, Imielinski MB, Cox JD, Wang JY, Ha SI, Keister BA, Chodosh LA (2002) Persistent parity-induced changes in growth factors, TGF-beta3, and differentiation in the rodent mammary gland. Mol Endocrinol 16:2034–2051

Goss PE (2004) Changing clinical practice: extending the benefits of adjuvant endocrine therapy in breast cancer. Semin Oncol 31:15–22

Henderson BE, Ross RK, Pike MC (1991) Toward the primary prevention of cancer. Science 254:1131–1138

Jordan VC, Labadidi MK, Langan-Fahey S (1991) Suppression of mouse mammary tumorigenesis by long-term tamoxifen therapy. J Natl Cancer Inst 83:492–496

Maffini MV, Soto AM, Calabro JM, Ucci AA, Sonnenschein C (2004) The stroma as a crucial target in rat mammary gland carcinogenesis. J Cell Sci 117:1495–1502

Medina D (2005) Mammary developmental fate and breast cancer risk. Endocr Relat Cancer 21:1–13

Medina D, Kittrell FS, Shepard A, Stephens LC, Jiang C, Lu J, Allred DC, McCarthy M, Ullrich RL (2002) Biological and genetic properties of the p53 null preneoplastic mammary epithelium. FASEB J 16:881–883

Nandi S, Guzman RC, Thordarson G, Rajkumar L (2005) Estrogen can prevent breast cancer by mimicking the protective effect of pregnancy. In: Li JJ, Li SA, Llombart-Bosch A (eds) Hormonal carcinogenesis IV. Springer, Berlin Heidelberg New York, pp 153–165

Osborne CK (1998) Tamoxifen in the treatment of breast cancer. N Engl J Med 339:1609–1618

Rajkumar L, Kittrell FS, Guzman RC, Brown PH, Nandi S, Medina D (2007) Hormone-induced protection of mammary tumorigenesis in genetically engineered mouse models. Breast Cancer Res 9:R12

Reddy M, Nguyen S, Farjamrad F, Laxminarayan S, Lakshmanaswamy R, Guzman R, Yang J, Nandi S (2002) Short-term hormone treatment with pregnancy levels of estradiol prevents mammary carcinogenesis by preventing promotion of carcinogen initiated cells. Proceedings of the American Association for Cancer Research Annual Meeting. American Association for Cancer Research 43:1964

Santen RJ (2003) Risk of breast cancer with progestins: critical assessment of current data. Steroids 68:953–964

Schedin P, Mitrenga T, McDaniel S, Kaeck KM (2004) Mammary ECM composition and function are altered by reproductive state. Mol Carcinogen 41:207–220

Sivaraman L, Stephens LC, Markaverich BM, Clark JA, Krnacik S, Conneely OM, O'Malley BW, Medina D (1998) Hormone-induced refractoriness to mammary carcinogenesis in Wistar-Furth rats. Carcinogenesis 19:1573–1581

Sivaraman L, Conneely OM, Medina D, O'Malley BW (2001) p53 is a potential mediator of pregnancy and hormone-induced resistance to mammary carcinogenesis. Proc Natl Acad Sci U S A 98:12379–12384

Stefanick ML, Anderson GL, Margolis KL, Hendrix SL, Rodabough RJ, Paskett ED, Lane DS, Hubbell FA, Assaf AR, Sarto GE, Schenken RS, Yasmeen S, et al (2006) Effects of conjugated equine estrogens on breast cancer and mammography screening in postmenopausal women with hysterectomy. JAMA 295:1647–1657

Swanson SM, Guzman RC, Collins G, Tafoya P, Thordarson G, Talamantes F, Nandi S (1995) Refractoriness to mammary carcinogenesis in the parous mouse is reversible by hormonal stimulation induced by pituitary isografts. Cancer Lett 90:171–181

Thordarson G, Jin E, Guzman RC, Swanson SM, Nandi S, Talamantes F (1995) Refractoriness to mammary tumorigenesis in parous rats: is it caused by persistent changes in the hormonal environment or permanent biochemical alterations in the mammary epithelia? Carcinogenesis 16:2847–2853

Thordarson G, Van Horn K, Guzman RC, Nandi S, Talamantes F (2001) Parous rats regain high susceptibility to chemically induced mammary cancer after treatment with various mammotropic hormones. Carcinogenesis 22:1027–1033

Thordarson G, Semaan S, Low C, Ochoa D, Leong H, Rajkumar L, Guzman RC, Nandi S, Talamantes F (2004a) Mammary tumorigenesis in growth hormone deficient spontaneous dwarf rats; effects of hormonal treatments. Breast Cancer Res Treat 87:277–290

Thordarson G, Slusher N, Leong H, Ochoa D, Rajkumar L, Guzman R, Nandi S, Talamantes F (2004b) Insulin-like growth factor (IGF)-I obliterates the pregnancy-associated protection against mammary carcinogenesis in rats: evidence that IGF-I enhances cancer progression through estrogen receptor-alpha activation via the mitogen-activated protein kinase pathway. Breast Cancer Res 6:R423–R436

Yang J, Yoshizawa K, Nandi S, Tsubura A (1999) Protective effects of pregnancy and lactation against N-methyl-N-nitrosourea-induced mammary carcinomas in female Lewis rats. Carcinogenesis 20:623–628

Ernst Schering Foundation Symposium Proceedings, Vol. 1, pp. 127–150
DOI 10.1007/2789_2007_059
© Springer-Verlag Berlin Heidelberg
Published Online: 17 December 2007

Estrogens, Progestins, and Risk of Breast Cancer

M.C. Pike[✉], A.H. Wu, D.V. Spicer, S. Lee, C.L. Pearce

Departments of Preventive Medicine and Medicine, University of Southern California/
Norris Comprehensive Cancer Center, 1441 Eastlake Avenue, Los Angeles, CA 90033,
USA
email: *mcpike@usc.edu*

Abstract. Obesity is associated with a decreased risk of breast cancer in premenopausal women but an increased risk in postmenopausal women, an effect that increases with time since menopause. Analysis of these effects of obesity shows that there is a ceiling to the carcinogenic effect of estrogen on the breast; increases in nonsex hormone-binding globulin-bound estradiol (non-SHBG bound E_2) exceeding approximately 10.2 pg/ml have no further effect on breast cancer risk; this ceiling is lower than the lowest level seen during the menstrual cycle. This suggests that the effects of menopausal estrogen therapy (ET) and menopausal estrogen–progestin therapy (EPT) on a woman's breast cancer risk will greatly depend on her body mass index (BMI; weight in kilograms/height in meters squared, kg/m^2) with the largest effects being in slender women. Epidemiological studies confirm this prediction. Our best estimates, per 5 years of use, of the effects of ET on breast cancer risk is a 30% increase in a woman with a BMI of 20 kg/m^2 decreasing to an 8% increase in a woman with a BMI of 30 kg/m^2; the equivalent figures for EPT are 50% and 26%. The analysis of the effects of estrogen also shows that even reducing the dose of estrogen in ET and EPT by as much as a half will have little or no effect on these risks. Reducing the progestin dose is likely to significantly reduce the risk of EPT: this is possible with an endometrial route of administration.

1 Introduction

1.1 Age-Incidence of Breast Cancer

The incidence of the common nonhormone-dependent adult cancers (stomach, colon, etc.) rises increasingly rapidly with age (Cook et al. 1969). This is illustrated in Fig. 1A for colorectal cancer in white women in the United States, 1969–1971 (Cutler and Young 1975); on a log–log scale the age-incidence curve of such a cancer is linear (Fig. 1B).

Breast cancer incidence also increases with age, but the rate of increase slows around menopause. This is illustrated in Fig. 2A for breast cancer in white women in the United States, 1969–1971 (Cutler and Young 1975), before mammographic screening and menopausal hormone therapy (HT) had distorted the picture. A log–log plot (Fig. 2B) clearly shows the major change that occurs around age 50. The important etiologic elements for breast cancer are thus sharply reduced following the menopause, i.e., the hormonal pattern of premenopausal women [cyclic production of relatively large amounts of estradiol (E_2) and progesterone (P_4)] causes a greater rate of increase in risk of breast cancer than the hormonal pattern of postmenopausal women (constant low E_2 and very low P_4). Other hormones may be involved but changes in E_2 and P_4 are sufficient to provide a comprehensive explanation of the major changes that occur as we demonstrate below.

preted as the same phenomenon as the observation that a late age at first birth is associated with a long-term increase in risk (see below).

2   Mathematical Modeling of Risk

The incidence at age t, $I(t)$, of the common nonhormone-dependent cancers (illustrated in Fig. 1A) rises as the kth power of age (Cook et al. 1969):

$$I(t) = a \times t^k$$

so that

$$\log[I(t)] = \log(a) + k \times \log(t)$$

and the rate of tissue aging (or more accurately, tissue cancer aging) at age t is constant, set at 1.0 per year of age, independent of t.

The breast cancer incidence curve fits into this form if age, t, is replaced by "breast tissue age," $b(t)$, so that:

$$I(t) = a \times [b(t)]^k$$

and

$$\log[I(t)] = \log(a) + k \times \log[b(t)]$$

The breast tissue aging rate at age t, $r(t)$, is the rate at which the underlying carcinogenic process is taking place, and $r(t)$ is summed from 0 to t to give $b(t)$ (Pike et al. 1983). $r(t)$ will be roughly equal to mitotic rate, or perhaps cell death and resulting replacement in some stem-cell compartment (Cairns 2002). The hypothesis is simply that hormones affect breast cancer incidence mainly through their effect on mitotic rates in the stem-cell compartment.

The breast tissue aging rates, i.e., the breast tissue $r(t)$s, that account in quantitative terms for the effects of menarche, menopause, births, and the incidence curve in Fig. 2, are shown in Fig. 3 (Pike et al. 1983). Breast tissue aging starts at menarche and continues at a constant rate up to first birth, then at a reduced rate until the start of the perimenopausal period, declining further gradually thereafter to a low postmenopausal rate. To account for a birth after around age 32 being associated with a breast cancer rate that is larger than that of nulliparous women, the model includes an increase in breast tissue aging during the pregnancy. Setting $r(t)$ from menarche to the first birth as 1.0 per year of age, the rate increases to approx. 3.2 per year during the year of the birth, and then falls to approx. 0.7 per year of age until the perimenopause, declining steadily thereafter to approx. 0.105 per year of age after menopause; the fitted exponent of time, k, is 4.5 (Pike et al. 1983). The fit of the model to the incidence curve is shown by the smooth curves in Fig. 2A and 2B, and the model provides an excellent description of the observed effects of ages at menarche, menopause, and births. The short-term increase in breast cancer risk after a birth can be equated to the observation that a late age at a birth is associated with a long-term increase in risk; effectively the benefit of a birth in reducing the breast tissue aging rate does not have a long enough time (before the postmenopausal drop in aging rate occurs) to compensate for the increased aging rate during the pregnancy. Rosner and colleagues (1994; Colditz and Rosner 2000) have extended this model to specifically incorporate the effects of second and later births.

3 Breast Cell Mitotic Activity

Breast epithelium cell proliferation varies over the menstrual cycle; proliferative activity over the luteal phase is roughly double the level over the follicular phase (Fig. 4; Goebelsmann and Mishell 1979; Anderson et al. 1982; Pike et al. 1993), so that approximately two-thirds of the breast cell proliferation in a premenopausal woman can be accounted for by an estrogen effect alone. Ovulatory cycling with the presence of progesterone is thus positively correlated with proliferative activity and is the likely explanation for why a more rapid establishment of regular cycles is associated with the greatest increase in breast cancer risk. Based on the fit shown in Fig. 2A and 2B, we calculate that the proliferative rate in the postmenopausal period for the average woman around 1970 was approximately one-fifth to one-quarter the rate during the follicular phase of the cycle [i.e., ratio of the $r(t)$s given above adjusted for phase of the menstrual cycle: $0.105/(2/3 \times 0.7) = 0.23$]; data on this are sparse, but this is in line with what has been found (Meyer and Connor 1982).

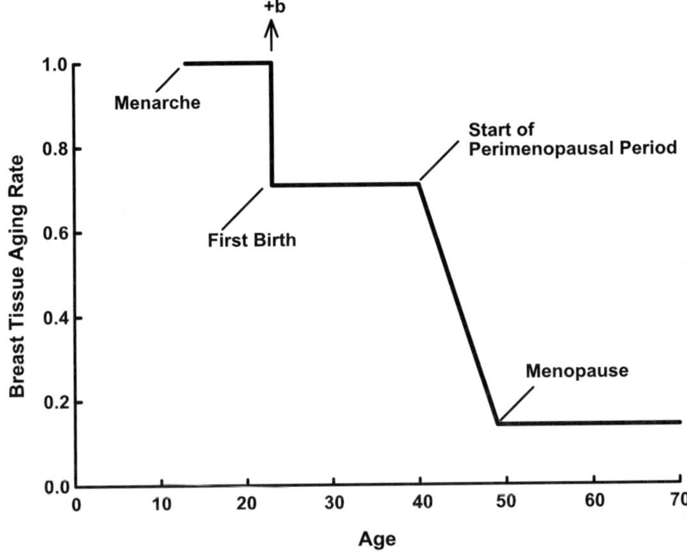

Fig. 3. Model of rate of breast tissue aging (Pike et al. 1983)

With this understanding of the age-incidence curve and the effects of menopause, menarche, and births in model terms, we are in a position to gain further quantitative insight into the effects of estrogens and progestins on risk of breast cancer.

4   Body Mass Index

Body mass index (BMI; weight in kg/height in meters squared, kg/m^2) has profound age-dependent effects on breast cancer risk. For older postmenopausal women, breast cancer risk increases with increasing BMI, while in premenopausal women increasing BMI is associated with a *decrease* in breast cancer risk (Kelsey and Bernstein 1996). Increased postmenopausal BMI is associated with increased serum estrogen levels (Endogenous Hormones and Breast Cancer Collaborative Group 2003a, 2003b), while a high premenopausal BMI is associated with increased

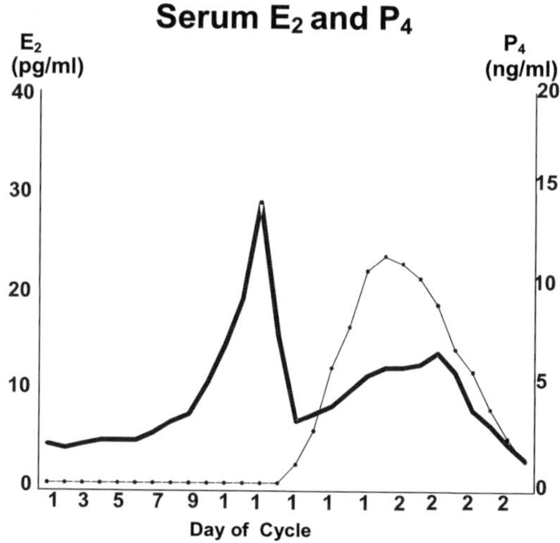

Breast Cell Proliferation

Relative Labelling Index

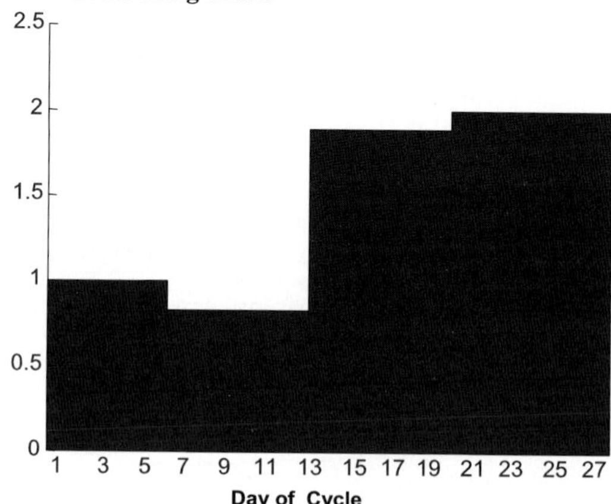

Fig. 4. Plasma E_2 and P_4, and breast tissue mitotic rate by day of menstrual cycle (Pike et al. 1993)

anovulation with decreased serum levels of both estrogen and progesterone (Hartz et al. 1979).

Table 1 shows, using the results of the pooled analysis of cohort studies (van den Brandt et al. 2000), the relative risks for a woman with a BMI of 30 kg/m^2 compared with a woman with a BMI of 20 kg/m^2 at the end of her premenopausal period at age 50 years and at a postmenopausal age of approx. 62 years some 12 years after a natural menopause at 50 years. The relative risk of 0.75 at age 50 becomes a relative risk of 1.20 at age 62; the associated breast tissue ages from the mathematical model described above are shown in the lower half of the table. The relative risk of 0.75 at age 50 implies a breast tissue age of approx. 27.88 years for the 30-kg/m^2 woman compared with a breast tissue age of approx. 29.73 years for the 20-kg/m^2 woman (assuming age at menarche = 12, age at first birth = 24, and the same parity of the average woman around 1970 for both women). If we estimate the breast tissue aging rate in the 20-kg/m^2 woman to be approx. 0.1, that is, somewhat less than the value found in fitting the model shown in Fig. 3, then the relative risk of 1.2 at age 62 implies a breast tissue age of approx. 32.20 years for the 30-kg/m^2 woman compared with a breast tissue age of approx. 30.93 years for the 20-kg/m^2 woman. Thus, the increase in breast tissue age per year, $r(t)$, in the postmenopausal period in

Table 1 Relative risks of breast cancer by BMI in premenopausal and postmenopausal women, and calculated associated breast tissue ages

	BMI	Premenopausal (Age 50)	Postmenopausal (Age 62)[*]
Relative risks:			
	$\sim 20\,\text{kg/m}^2$	1	1
	$\sim 30\,\text{kg/m}^2$	0.75	1.20
Breast tissue age:			
	$\sim 20\,\text{kg/m}^2$	29.73 yrs	30.93 yrs
	$\sim 30\,\text{kg/m}^2$	27.88 yrs	32.20 yrs

[*]Age at menopause 50 years

the 30-kg/m^2 woman is the difference between her breast tissue ages at age 62 and age 50, namely, $(32.20 - 27.88) = 4.32$ years, divided by 12 (the number of years between 50 and 62), so that, $r(t) = 4.32/12 = 0.36$.

Equating the $r(t)$ values of 0.1 and 0.36 for the 20-kg/m^2 and 30-kg/m^2 women, respectively, with the "effective estrogen" level, the relative levels of effective estrogen will be $0.36/0.1 = 3.6$. With no threshold, this 3.6 is simply the ratio of effective estrogen levels, but with a threshold value for effective estrogen, the ratio of effective estrogen will be lower and may be considerably lower. If the effective estrogen is equated to serum nonsex hormone-binding globulin-bound E_2 (non-SHBG bound E_2), i.e., approx. 8.4 pg/ml and approx. 4.0 pg/ml for a woman with a BMI of 30 kg/m^2 and a woman with a BMI of 20 kg/m^2, respectively [calculated with the mass action approach of Södergard and colleagues (1982) with the association constants and E_2 and SHBG values used by the Endogenous Hormones and Breast Cancer Collaborative Group (2003a; 2003b; Dunn et al. 1981)], we find that the ratio of effective estrogen is $8.4/4.0 = 2.1$. A threshold value of non-SHBG bound E_2 of 2.31 pg/ml converts this to a 3.6 ratio, i.e., $(8.4–2.31)/(4.0–2.31) = 3.6$.

Approximately two-thirds of the breast cell proliferation in a premenopausal woman can be accounted for by an estrogen effect alone (Fig. 4). The breast tissue aging rate, $r(t)$, of the parous premenopausal woman is approximately 0.7, so we estimate that the effect due solely to estrogen is $0.7 \times 2/3 = 0.47$, and the effect of the estrogen level of the 30-kg/m^2 woman is $0.36/0.47 = 77\%$ that of the effect of follicular phase estrogen. If we continue to equate effective estrogen with non-SHBG bound E_2, the non-SHBG bound E_2 level of the postmenopausal woman with a BMI of 30 kg/m^2 is approx. 8.4 pg/ml, so that, taking into account the non-SHBG bound E_2 threshold of 2.31 pg/ml, the effective ceiling for non-SHBG bound E_2 effect is unlikely to be greater than $(8.4 - 2.31)/0.77 + 2.31 = 10.2$ pg/ml.

5 Predicted Effects of Menopausal Hormone Therapy

Menopausal estrogen therapy (ET) at a conjugated equine estrogen (CEE) dose of 0.625 mg/day, the most commonly used dose in the

United States until very recently, is roughly equivalent to a 50-μg E_2 transdermal patch. A 50-μg E_2 patch increases steady-state E_2 levels by approx. 30 pg/ml (Schiff et al. 1982; Powers et al. 1985; Selby and Peacock 1986) and has little effect on SHBG (Nachtigall et al. 2000; Ropponen et al. 2005), so that non-SHBG bound E_2 will increase from approx. 4.0 pg/ml to approx. 18.4 pg/ml in a 20-kg/m^2 woman and from approx. 8.4 pg/ml to approx. 27.4 pg/ml in a 30-kg/m^2 woman (calculated as described above). The steady-state non-SHBG bound E_2 level of all women on a 50-μg E_2 patch is thus well above the ceiling level of 10.2 pg/ml estimated above. The effect of the 0.625-mg/day dose of CEE is thus also likely to be well above the estrogen ceiling dose.

A postmenopausal woman on ET is thus predicted to have a breast tissue aging rate of approx. 0.47 (two-thirds of 0.7) *independent* of her BMI assuming she takes the ET every day (Fig. 5). The predicted breast

Table 2 Predicted relative risks (RR$_5$s) of breast cancer at age 55 years (age at menopause 50 years) after 5 years of ET or EPT use by BMI

BMI		No HT	ET	ccEPT2.5[a]	ccEPT10[b]	sEPT10[c]
20 kg/m^2	Breast tissue age	30.23[d]	32.06	33.07	36.10	33.50
	RR$_5$[g]	1.00	1.30	1.50	2.22	1.59
30 kg/m^2	Breast tissue age	29.68[e]	30.22	31.23	34.26	31.66
	RR$_5$[f]	0.92	1.00	1.16	1.76	1.23
	RR$_5$[g]	1.00	1.08	1.26	1.91	1.34

[a]Continuous-combined EPT with MPA at 2.5 mg/day. [b]Continuous-combined EPT with MPA at 10 mg/day. [c]Sequential EPT with MPA at 10 mg/day for 10 days per 28-day cycle. [d]This figure is calculated as the sum of the 29.73 breast tissue age shown in Table 1 at age 50 plus 5 years of the breast tissue aging rate of 0.1 in the postmenopausal period. [e]This figure is calculated as the sum of the 27.88 breast tissue age shown in Table 1 at age 50 plus 5 years of the breast tissue aging rate of 0.36 in the postmenopausal period. [f]Relative to a 20-kg/m^2 woman not using HT. [g]Relative to a 30-kg/m^2 woman not using HT

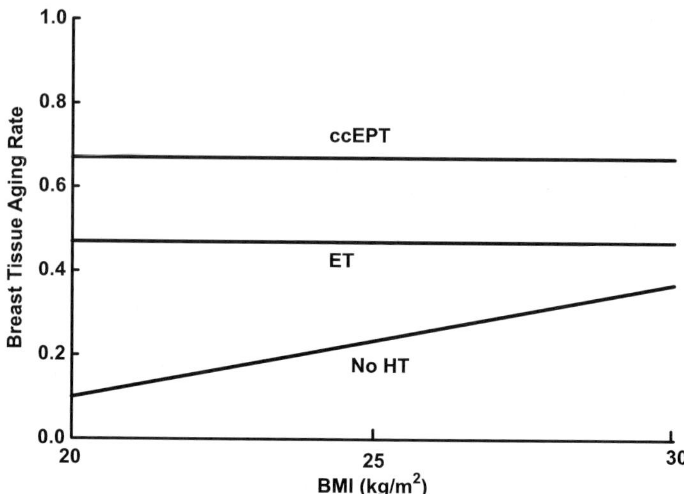

Fig. 5. Breast tissue aging rate in postmenopausal women by BMI. *ccEPT*, continuous-combined estrogen–progestin therapy at a medroxyprogesterone acetate (MPA) dose of 2.5 mg/day; *ET*, estrogen therapy; *HT*, ET or EPT

cancer consequences of this for 5 years of ET use are shown in Table 2 and Fig. 6 for a woman aged 55 years whose age at menopause was 50. The predicted effect of ET use in a woman of 20 kg/m^2 is to increase her breast cancer risk by 30% after 5 years of use. The predicted effect of ET use in a woman of 30 kg/m^2 is to increase her breast cancer risk by only 8% after 5 years of use.

The mathematical model can also be used predict what the effect of menopausal estrogen–progestin therapy (EPT) will be. We illustrate this for continuous-combined EPT as used in the United States. The area under the curve (AUC) of P_4 during a normal menstrual cycle is approximately 97 ng/ml (Fig. 4). The continuous-combined progestin used in the United States was an oral dose of CEE of 2.5 mg/day of medroxyprogesterone acetate (MPA). An oral MPA dose of 2.5 mg/day is estimated to have the same progestin effect as 125 mg/day of oral P_4, and this dose of P_4 is estimated to produce a serum P_4 level equivalent to a steady-state level of 3.0 ng/ml (Stanczyk 2002); leading to an

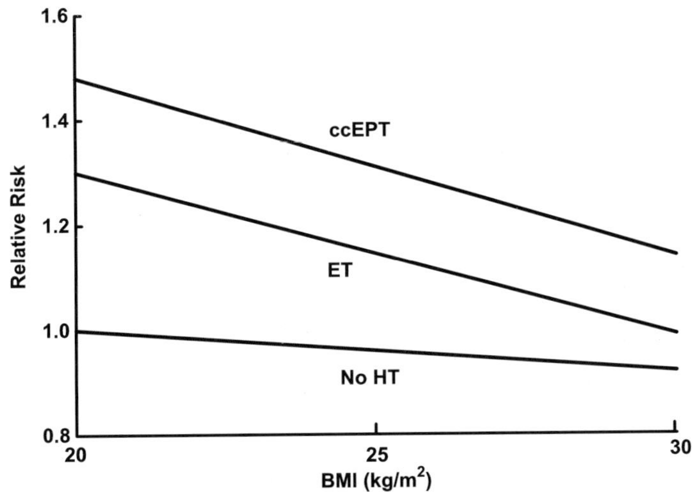

Fig. 6. Relative risks of breast cancer at age 55 years (age at menopause 50 years) after 5 years of HT use by BMI. *ccEPT*, continuous-combined estrogen-progestin therapy at an MPA dose of 2.5 mg/day; *ET*, estrogen therapy; *HT*, ET or EPT

AUC of 84 ng/ml per 28-day cycle. On a straightforward linear model of P_4 action one would therefore predict a 0.87 (= 84/97) progestin effect compared with that seen during a menstrual cycle. The contribution to the breast cell aging rate from the progesterone during a menstrual cycle is roughly one-third of the total effect, that is, 0.7/3=0.23, so that the MPA at 2.5 mg/day is estimated to add (0.23×0.87)=0.20 to the rate associated with the estrogen component of EPT, namely, 0.47, giving a total breast cell aging rate of 0.67. The predicted breast tissue aging rate for use of this continuous-combined EPT is shown in Fig. 5—the rates are again *independent* of BMI. The predicted effects of such a regimen for 5 years on breast cancer incidence are also shown in Table 2 and in Fig. 6 for a woman aged 55 years whose age at menopause was 50 years. It is worth noting here too that the predicted effect of such EPT use in a woman of 20 kg/m² is to increase her breast cancer risk by 50% after 5 years of use. The predicted effect of such EPT use in a woman of 30 kg/m² is to increase her breast cancer risk by 26%, roughly half the relative risk seen in the 20-kg/m^2 woman but still a significant added risk. (Note: The relationship of BMI to breast cancer risk shown in Table 2 and Fig. 6 is very slight in contrast to the effect of BMI shown in Table 1. This is because the effect of BMI on risk becomes greater and greater with increasing time since menopause, and the results shown in Table 2 and Fig. 6 refer to approx. 5 years after menopause, whereas the results shown in Table 1 refer to approx. 12 years after menopause. Immediately after menopause there is still a negative relationship between weight and risk as a carry-over from the reduced risk with increased weight seen in premenopausal women. At older ages the effect of increasing BMI will be greater than that noted in Table 1.)

Table 2 also shows the predicted effect of 5 years of use of other EPT regimens. The most commonly used sequential EPT regimen in the United States was 10 days of MPA at a dose of 10 mg/day. The predicted effect of this regimen is a 59% and a 34% increase in breast cancer risk in a woman of 20 kg/m^2 and 30 kg/m^2, respectively, slightly greater than the predicted effect of the 2.5-mg/day MPA continuous-combined regimen.

These predicted effects are robust but not precise. A significant number of assumptions have had to be made, for example, that the relative risks associated with obesity refer to women aged 50 and 62 respectively, and we have had to use hormone level figures from different studies, whereas one actually needs the measurements to have been made at the same time using the same methods. However, the overall picture is quite clear and we shall see in the next section that these predicted effects are precisely what are seen in epidemiologic studies.

6   Epidemiologic Studies of Estrogen Therapy and Estrogen–Progestin Therapy

6.1   Estrogen Therapy

Epidemiologic studies show that menopausal ET causes an approximately 2% increase in breast cancer risk per year of use (Collaborative Group on Hormonal Factors in Breast Cancer 1997); i.e., a 10% increase in risk after 5 years of use or a cumulative increase in risk for 5 years of use of approximately half this amount. The increased risk is much more evident in slender women and little if any effect is seen in obese women (Table 3; Collaborative Group on Hormonal Factors in Breast Cancer 1997; Schairer et al. 2000; Million Women Study Collaborators 2003; Beral et al. 2005). The model as further developed in the section on BMI above predicts precisely these results, and the few epidemiologic study results published are in good agreement with the predictions shown in Table 2 and Fig. 6.

The above paragraph's discussion of ET is based on an overview of all epidemiologic studies of the effect of ET on breast cancer risk. However, the Women's Health Initiative (WHI) randomized trial in hysterectomized women found a decrease in risk with use of ET (Women's Health Initiative 2004). The lack of effect of ET in the WHI trial is difficult to reconcile with at least three other observations: (1) increased postmenopausal serum estrogen levels are associated with increased risk; risk increases two- to three-fold between the extreme quintiles of E_2, estrone (E_1) and E_1 sulfate (Endogenous Hormones and Breast Cancer Collaborative Group 2003a, 2003b); (2) increasing postmenopausal

Table 3 Observed relative risks (RR) of breast cancer from HT use by BMI

Study	HRT type	BMI (kg/m^2)	RR
Million Women	ET	<25	1.36
Study Collaborators		25–29	1.14
(2003; Beral et al. 2005)[a]		30+	0.99
	EPT	<25	2.31
		25+	1.78
Schairer et al. (2000)[b]	ET	<= 24.4	1.03
		> 24.4	0.99
	EPT	<= 24.4	1.12
		> 24.4	1.04

[a]RR in current users over 2.6 years of follow-up
[b]RR per year of use

weight increases breast cancer risk and is largely explained by the association of increased weight with increased serum estrogen levels (Endogenous Hormones and Breast Cancer Collaborative Group 2003a, 2003b); and (3) the risk of contralateral breast cancer is sharply reduced by treating the original cancer with an aromatase inhibitor that drastically reduces estrogen levels. The evidence considered as a whole shows that ET increases breast cancer risk, and that the effect is greatest in slender women and difficult to see in women with a BMI over 30 kg/m^2.

6.2   Estrogen–Progestin Therapy

The overview of HT studies by the Collaborative Group on Hormonal Factors in Breast Cancer (1997) contained very little information on the effects of EPT. Since that time a number of studies have provided information on the breast cancer risk from EPT use in a quantitative assessment that adjusted for age at menopause, a critical factor in assessing HT use and breast cancer risk, and provided information on risk by duration of EPT use. The results from these studies and from the Collaborative Group on Hormonal Factors in Breast Cancer (1997) pooled analysis were used to obtain an overall assessment of EPT and breast cancer risk by Lee et al. (2005).

Lee et al. (2005) took special measures to permit the results from the WHI randomized trial (Chlebowski et al. 2003), cohort studies, and case-control studies to be expressed in the same relative risk terms. First, the WHI randomized trial found an average relative risk of breast cancer of 1.24 for EPT use after an average of 5.6 years of follow-up. This 1.24 figure converts to a relative risk after 5 years of use, RR_5, of 1.47. Second, for cohort studies, the true duration of EPT use is underestimated in current hormone users. This is because EPT use is assessed at baseline but continues for an unknown proportion of individuals for at least some further period until last follow-up time, so that an additional duration of use should be added for current hormone users in the cohort studies considered. Lee et al. (2005) made the conservative assumption that current users of EPT remained users during follow-up; this approach slightly underestimates the effect of EPT use, because some women will have stopped use before the last follow-up date in the studies.

The overall summary of the studies (Table 4) showed a weighted average relative risk at the end of 5 years of use, RR_5, of 1.44. The RR_5 for the studies from the United States was 1.29, and for the Scandinavian studies was 1.53, a difference that was highly statistically significant. In the studies from the United States, the continuous-combined regimen was associated with a slightly lower risk, namely a 20% increased risk after 5 years of continuous-combined use compared to a 32% increase with sequential therapy, but in the Scandinavian studies the opposite was observed with the continuous-combined regimens being associated with an 88% increased risk compared to a 40% increase with sequential use. These differences are explained by a combination of the different continuous-combined progestin doses in the United States and Scandinavia and the much greater BMI of women from the United States.

Table 4 Observed relative risks of breast cancer after 5 years of use (RR_5s) of estrogen-progestin therapy (Lee et al. 2005)

Studies	RR_5 (95% CI)
All studies	1.44 (1.40–1.48)
U.S. studies	1.29 (1.19–1.39)
Scandinavian studies	1.53 (1.37–1.72)

CI, confidence interval

Table 5 Observed relative risks of breastcancer after 5 years of use (RR_5s) of estrogen-progestin therapy by progestin schedule (Lee et al. 2005)

Studies	Sequential RR_5 (95% CI)	Continuous-combined RR_5 (95% CI)
All studies	1.53 (1.47 – 1.60)	1.63 (1.55 – 1.72)
U.S. studies	1.32 (1.11 – 1.56)	1.20 (1.01 – 1.44)
Scandinavian studies	1.40 (1.19 – 1.64)	1.88 (1.61 – 2.21)

CI, confidence interval

In the United States, the most common form of sequential EPT provided 10 mg/day of MPA for 10 days per 28-day cycle, whereas subjects assigned to receive continuous-combined EPT were given 2.5 mg/day of MPA every day. With these regimens the total doses for sequential and continuous-combined are 100 mg and 70 mg, respectively, per cycle. In contrast, in Scandinavia the total dose of the progestin is much higher with continuous-combined than with sequential EPT, at least for two commonly prescribed regimens using norethisterone acetate (NETA). In these regimens, the same daily dose of NETA, 1 mg, is used with both the sequential and the continuous-combined EPT, so that the total NETA dose per cycle is roughly 10 mg and 28 mg, respectively, and the increased risk from the continuous-combined regimen in Scandinavia is much greater than the increased risk seen with the sequential regimen (Table 5; an 88% and 40% increase, respectively). Some of the greater effects seen within the sequential regimens in Scandinavia compared to the United States (Table 5; 40% and 32% increase respectively) may be due to a greater effect of NETA at 1 mg/day compared to 10 mg/day of MPA, although these doses are generally considered equivalent based on their endometrial effects (Stanczyk 2002). Most of the remaining excess in the Scandinavian studies relative to the studies from the United States is explained by a greater relative effect of EPT use on breast cancer risk among leaner women (Table 3). Women in the studies from the United States were much heavier than the women in the European studies (Beral et al. 2005).

7   Conclusions

ET use is associated with a significantly increased risk of breast cancer, a risk that is especially significant in slender women due to their much lower endogenous levels of estrogen and to the fact that the ceiling of estrogen effect on the breast is quite low. The ceiling level is not much higher than that found in a woman with a BMI of 30 kg/m^2, so that the effect of ET on breast cancer risk in such women or heavier women is small. Even if the estrogen dose in ET is reduced by 50% (transdermal patch of 25 μg E_2 or equivalent) the non-SHBG bound E_2 will likely still be slightly greater than the ceiling level of 10.2 pg/ml so that no reduction in the breast cancer risk from ET or EPT use will be seen.

EPT use is associated with a much greater increased risk of breast cancer than ET use, and this increase, although greater in slender women, is still very evident in heavier women due to the fact that progestin appears to act independently of estrogen on the breast without the estrogen ceiling blunting the progestin effect. Reducing the progestin dose is likely to significantly reduce the risk; this is possible with an endometrial route of administration (Shoupe et al. 1991; Varila et al. 2001).

Since the publication of the Lee et al. (2005) review of EPT and breast cancer, a small French study has reported no increase in breast cancer risk if micronized progesterone is used as the progestin in EPT (Fournier et al. 2005), and an experimental study in macaques also showed no effect of micronized progesterone on breast cell proliferation (Wood et al. 2007). This is, of course, potentially an extremely important finding, and if confirmed this will mean that not only will EPT use need to be reevaluated but it will also mean that the use of micronized progesterone as the progestin in oral contraceptives will also need to be evaluated for a possible protective effect against breast cancer. However, mammographic density was found to increase with micronized progesterone as the EPT progestin in a randomized trial (Greendale et al. 2003) so that the situation is currently unclear. This issue needs much further study. We are currently conducting a study measuring breast-cell proliferation in women on such a progesterone-based EPT. If progesterone is truly not mitogenic in the human breast then we need to find the factor or factors that could be producing the luteal phase increase in breast cell proliferation; it cannot be estrogen as the level of estrogen exceeds the mitogenic ceiling throughout the menstrual cycle.

Acknowledgements. This work was supported by a Department of Defense Congressionally Directed Breast Cancer Program Grant BC 044808; by the USC/Norris Comprehensive Cancer Center Core Grant P30 CA14089; and by generously donated funds from the endowment established by Flora L. Thornton for the Chair of Preventive Medicine at the USC/Norris Comprehensive Cancer Center. The funding sources had no role in this report.

References

Anderson TJ, Ferguson DJ, Raab GM (1982) Cell turnover in the "resting" human breast: influence of parity, contraceptive pill, age and literality. Br J Cancer 46:376–382

Beral V, Reeves G, Banks E (2005) Current evidence about the effect of hormone replacement therapy on the incidence of major conditions in postmenopausal women. Br J Obstet Gynaecol 112:692–695

Cairns J (2002) Somatic stem cells and the kinetics of mutagenesis and carcinogenesis. Proc Natl Acad Sci USA 99:10567–10570

Chlebowski RT, Hendrix SL, Langer RD et al. (2003) Influence of estrogen plus progestin on breast cancer and mammography in healthy postmenopausal women: the Women's Health Initiative Randomized Trial. JAMA 289:3243–3253

Colditz GA, Rosner B (2000) Cumulative risk of breast cancer to age 70 years according to risk factor status: data from the Nurses' Health Study. Am J Epidemiol 152:950–964

Collaborative Group on Hormonal Factors in Breast Cancer (1997) Breast cancer and hormone replacement therapy: collaborative reanalysis of data from 51 epidemiological studies of 52,705 women with breast cancer and 108,411 women without breast cancer. Lancet 350:1047–1059

Cook PJ, Doll R, Fellingham SA (1969) A mathematical model for the age distribution of cancer in man. Cancer 4:93–112

Cutler SY, Young JL (1975) Third national cancer survey: incidence data. National Cancer Institute monograph No. 41. National Cancer Institute, Washington DC

Dunn JF, Nisula BC, Rodbard D (1981) Transport of steroid hormones: binding of 21 endogenous steroids to both testosterone-binding globulin and corticosteroid-binding globulin in human plasma. J Clin Endocrinol Metab 53:58–68

Endogenous Hormones and Breast Cancer Collaborative Group (2003a) Free estradiol and breast cancer risk in postmenopausal women: comparison of measured and calculated values. Cancer Epidemiol Biomarkers Prev 12:1457–1461

Endogenous Hormones and Breast Cancer Collaborative Group (2003b) Body mass index, serum sex hormones, and breast cancer risk in postmenopausal women. J Natl Cancer Inst 95:1218–1226

Fournier A, Berrino F, Riboli E et al. (2005) Breast cancer risk in relation to different types of hormone replacement therapy in the E3N-EPIC cohort. Int J Cancer 114:448–454

Goebelsmann U, Mishell DR (1979) The menstrual cycle. In: Mishell DR, Davajan V (eds) Reproductive endocrinology, infertility and contraception. FA Davis, Philadelphia, pp 67–89

Greendale GA, Reboussin BA, Slone S et al. (2003) Postmenopausal hormone therapy and change in mammographic density. J Natl Cancer Inst 95:30–37

Hartz AJ, Barboriak TN, Wong A et al. (1979) The association of obesity with infertility and related menstrual abnormalities in women. Int J Obes 3:57–73

Henderson BE, Ross RK, Judd HL et al. (1985) Do regular ovulatory cycles increase breast cancer risk? Cancer 56:1206–1208

Kelsey JL, Bernstein L (1996) Epidemiology and prevention of breast cancer. Annu Rev Public Health 17:47–67

Lee SA, Ross RK, Pike MC (2005) An overview of menopausal oestrogen-progestin hormone therapy and breast cancer risk. Br J Cancer 92:2049–2058

MacMahon B, Cole P, Brown J (1973) Etiology of human breast cancer: a review. J Natl Cancer Inst 50:21–42

Meyer JS, Connor RE (1982) Cell proliferation in fibrocystic disease and post-menopause breast ducts measured by thymidine labcling. Cancer 50:746–750

Million Women Study Collaborators (2003) Breast cancer and hormone-replacement therapy in the Million Women Study. Lancet 362:419–427

Nachtigall LE, Raju U, Banerjee S et al. (2000) Serum estradiol profiles in postmenopausal women undergoing three common estrogen replacement therapies: associations with sex hormone-binding globulin, estradiol, and estrone levels. Menopause 7:243–250

Pike MC, Krailo MD, Henderson BE et al. (1983) "Hormonal" risk factors, "breast tissue age" and the age-incidence of breast cancer. Nature 303:767–770

Pike MC, Spicer DV, Dahmoush L, Press MF (1993) Estrogens, progestogens, normal breast cell proliferation, and breast cancer risk. Epidemiol Rev 15:17–35

Powers MS, Schennkel L, Darley PE et al. (1985) Pharmacokinetics and pharmacodynamics of transdermal dosage forms of 17-β estradiol: comparison with conventional oral estrogens used for hormone replacement. Am J Obstet Gynecol 152:1099–1106

Ropponen A, Aittomaki K, Vihma V et al. (2005) Effects of oral estradiol administration on levels of sex hormone-binding globulin in postmenopausal women with and without a history of intrahepatic cholestasis of pregnancy. J Clin Endocrinol Metab 90:3431–3434

Rosner B, Colditz GA, Willett WC (1994) Reproductive risk factors in a prospective study of breast cancer: The Nurses' Health Study. Am J Epidemiol 139:819–835

Schairer C, Lubin J, Troisi R et al. (2000) Menopausal estrogen and estrogen-progestin replacement therapy and breast cancer risk. JAMA 283:485–491

Schiff I, Sela HK, Cramer D et al. (1982) Endometrial hyperplasia in women on cyclic or continuous estrogen regimens. Fertil Steril 37:79–82

Selby PL, Peacock M (1986) Dose dependent response of symptoms, pituitary, and bone to transdermal estrogen in postmenopausal women. BMJ 293:1337–1339

Shoupe D, Meme D, Mezrow G, Lobo RA (1991) Prevention of endometrial hyperplasia in postmenopausal women with intrauterine progesterone. N Engl J Med 325:1811–1812

Södergard R, Backstrom T, Shanbhag V, Carstensen H (1982) Calculation of free and bound fractions of testosterone and estradiol-17 beta to human plasma proteins at body temperature. J Steroid Biochem 16:801–810

Stanczyk FZ (2002) Pharmacokinetics and potency of progestins used for hormone replacement therapy and contraception. Rev Endocr Metab Disord 3:211–224

Trichopoulos D, MacMahon B, Cole P (1972) Menopause and breast cancer risk. J Natl Cancer Inst 48:605–613

Trichopoulos D, Hsieh CC, MacMahon B et al. (1983) Age at any birth and breast cancer risk. Int J Cancer 31:701–704

van den Brandt P, Spiegelman D, Yaun SS et al. (2000) Pooled analysis of prospective cohort studies on height, weight, and breast cancer risk. Am J Epidemiol 152:514–527

Varila E, Wahlstrom T, Rauramo I (2001) A 5-year follow-up study on the use of a levonorgestrel intrauterine system in women receiving hormone replacement therapy. Fertil Steril 76:969–973

Wohlfahrt J, Melbye M (2001) Age at any birth is associated with breast cancer risk. Epidemiology 12:68–73

Women's Health Initiative (2004) Effects of conjugated equine estrogen in postmenopausal women with hysterectomy. JAMA 291:1701–1712

Wood CE, Register TC, Lees CJ et al. (2007) Effects of estradiol with micronized progesterone or medroxyprogesterone acetate on risk markers for breast cancer in postmenopausal monkeys. Breast Cancer Res Treat 101:125–134

Ernst Schering Foundation Symposium Proceedings, Vol. 1, pp. 151–170
DOI 10.1007/2789_2008_077
© Springer-Verlag Berlin Heidelberg
Published Online: 29 February 2008

In Vivo Characterization of Progestins with Reduced Non-genomic Activity In Vitro

C. Otto(✉), B. Rohde-Schulz, G. Schwarz, I. Fuchs, M. Klewer,
H. Altmann, K.-H. Fritzemeier

TRG Women's Healthcare, Bayer Schering Pharma AG, Müllerstr. 178, 13342 Berlin,
Germany
email: *christiane.otto@bayerhealthcare.com*

Abstract. Postmenopausal women that still have an uterus and suffer from
hot flushes are treated with combinations of estrogens and progestins. Whereas
estrogens are indispensable for treating postmenopausal symptoms, progestins
are added to counteract the proliferative activity of estrogens on uterine epithelial cells. However, in the mammary gland, progestins, given together with
estrogens, stimulate the proliferation of mammary epithelial cells. Therefore,
progestins with reduced proliferative activity in the mammary gland would be
of advantage for hormone therapy of postmenopausal women. In order to identify progestins with better tissue-selectivity, we exploited the activation of different signal transduction pathways by the classical progesterone receptor. We
demonstrated that progestins with reduced non-genomic versus genomic activ-

ity in vitro show a better dissociation of uterine versus mammary gland effects in vivo than medroxyprogesterone acetate (MPA), a synthetic progestin that is widely used in hormone therapy.

1   Introduction

The progesterone receptor (PR) can activate genomic as well as rapid, non-genomic effects. In the classical genomic pathway, ligand-bound PR translocates into the nucleus, binds to progesterone-responsive elements (PREs) in the promoters of target genes, and stimulates gene transcription. Genomic effects have an onset in the range of minutes to hours and are sensitive towards inhibitors of transcription and translation. In contrast, rapid, non-genomic progesterone effects exhibit an onset within seconds to minutes and are insensitive to inhibitors of transcription and translation (Falkenstein et al. 2000). Rapid, non-genomic effects involve the activation of cytoplasmic signal transduction cascades. Direct activation of the src/p21ras/ERK signal transduction pathway (Migliaccio et al. 1998a) or the PI3K/Akt signal transduction pathway (Bagowski et al. 2001) by the classical PR has been described. Non-genomic activation of the PI3K/Akt pathway by progestins plays an important role during amphibian oocyte maturation (Bagowski et al. 2001). Non-genomic activation of the src/p21ras/ERK pathway by the PR seems to be involved in human breast cancer cell proliferation (Castoria et al. 1999). In breast cancer cells, src kinase activity is stimulated 2–5 min after progesterone treatment and leads to phosphorylation of Shc. Phosphorylated Shc promotes the activation of the ras/raf/MAP kinase cascade leading to induction of cyclin D1 and hence cellular proliferation (Castoria et al. 1999). There are two hypotheses regarding the activation of src by progestins: in GST-pulldown experiments, a direct interaction of PR with src has been demonstrated (Boonyaratanakornkit et al. 2001). In contrast, a different hypothesis postulates that the estrogen receptor (ER) and the PR form an inactive complex. Upon hormonal stimulation with either E2 or progesterone, ER is set free and then interacts with src, leading to cellular proliferation of breast cancer cells (Migliaccio et al. 1998b). In this latter case, the activation of src by PR would occur in an indirect manner.

Interestingly, mice deficient for cyclin D1 or progesterone receptor B (PRB) show the same mammary gland phenotype, i.e. reduced ductal sidebranching and lobuloalveolar development during pregnancy. Since the cyclin D1 promoter does not contain any PREs, we speculated that, also in vivo, non-genomic induction of cyclin D1 by progestins might play a role in normal mammary epithelial cell proliferation. Therefore, progestins with reduced nongenomic activity may have less proliferative activity in the mammary gland.

Synthetic progestins are widely used in hormone therapy in post-menopausal women that still have a uterus. Progestins counteract the proliferative effects of estradiol in the endometrium and thus prevent the generation of endometrial carcinomas by estrogen-only treatment (Hulka et al. 1982). However, the opposite occurs in the mammary gland epithelium. Several studies have demonstrated that progestins, if added to estradiol, lead to enhanced proliferation of mammary epithelial cells in several animal models (Said et al. 1997) and human breast tissue (Hofseth et al. 1999).

It is well-established that cell proliferation in response to progesterone and estradiol is regulated in a paracrine manner in the normal mammary gland, whereas it is regulated in an autocrine way in breast cancer cells. One suggested paracrine mechanism stimulating epithelial cell proliferation in the normal mammary gland involves receptor activator of NF-κB ligand (RANKL) upregulation in ER/PR-positive cells. RANKL binds to RANK in neighbouring cells (that are devoid of ER and PR) and mediates the activation of nuclear factor κB, which leads to enhanced cyclin D1 expression and cellular proliferation (Mulac-Jericevic et al. 2003). In contrast, breast cancer cells are often positive for ER and PR and thus show a direct proliferative response after stimulation with estradiol and progesterone (Anderson and Clarke 2004). In light of these results, our hypothesis that also in the normal mammary gland non-genomic activation of the PR might contribute to epithelial cell proliferation does not seem very convincing, since the PR is not expressed in normal proliferating cells. Assuming that premalignant mammary epithelial cells may have lost the paracrine regulation of cellular proliferation and may have switched to autocrine-controlled proliferation, we decided to search for progestins that stimulate non-genomic effects to reduced extent. Such progestins might be of advantage when used in combined hormone therapy since they would stimulate proliferation of such premalignant cells to a reduced extent. Independent from this hypothesis, we were interested in identifying progestins that lead to reduced stimulation of non-genomic effects and in analysing how this reduced non-genomic activity would translate into the in vivo situation.

2 Establishment of In Vitro Assays Monitoring Genomic and Non-genomic Progesterone Receptor Activity

Before developing in vitro assay systems that would allow the analysis of non-genomic progestin effects, we tried to reproduce the main findings from the literature. To monitor the induction of cyclin D1 by progestins in vitro, we used T47D cells that were serum-starved for 48 h and then treated with the synthetic progestin R5020 (10 nM) for different periods. Cellular extracts were prepared and analysed using Western blot (Fig. 1a). An antibody against β-tubulin served as the loading control. Cyclin D1 was rapidly induced by R5020, reaching a maximum between 4 h and 8 h. To further analyse the mechanism of cyclin D1 induction by progestins, we preincubated serum-starved T47D cells for 30 min with different kinase inhibitors before stimulation with R5020 was performed for 4 h (Fig. 1b). As inhibitors we used the mitogen-activated protein kinase kinase 1 (MEK1) inhibitor U0126 (20 nM), the src kinase inhibitor PP2 (15 µM), PP3 (15 µM) as inactive variant of PP2, the PI3K inhibitor Ly29402 (50 µM), the MAPK inhibitor PD98059 (15 µM) and the PR antagonist RU486 (10 µM). As demonstrated in Fig. 1b, the induction of cyclin D1 by progestins was PR-dependent and required the activation of non-genomic signal transduction pathways. Cyclin D1 induction by R5020 was completely dependent on PI3K activation and partly dependent on src kinase activation (Fig. 1b). In contrast to published data, mitogen-activated protein kinase (MAPK) activation did not play a major role in the induction of cyclin D1 by progestins in T47D cells (Boonyaratanakornkit et al. 2007).

To further analyse the interaction between src kinase and PRB, we performed GST-pulldown experiments using radiolabelled PR and several src kinase GST fusion proteins of different lengths. In line with data published previously (Boonyaratanakornkit et al. 2001), we were able to

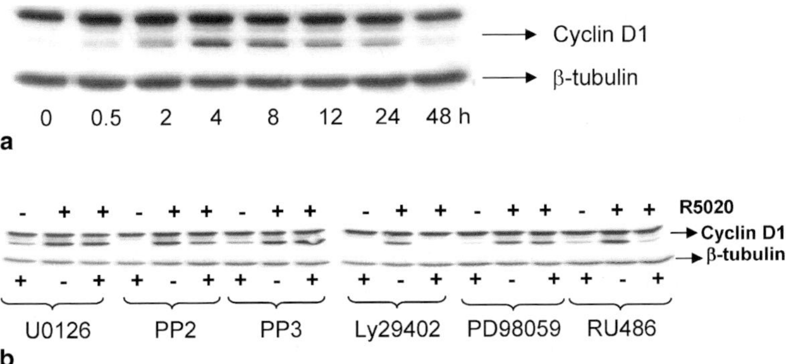

Fig. 1a, b. Cyclin D1 is induced non-genomically by progestins in T47D cells. T47D cells were serum-starved for 48 h and subsequently treated with 10 nM R5020 for the indicated times. Cyclin D1 expression was analysed by Western blot; β-tubulin served as the loading control. Induction of cyclin D1 starts 30 min after hormonal stimulation and reaches a maximum between 4 h and 8 h (**a**). To analyse the mechanisms leading to cyclin D1 induction by progestins, T47D cells were serum-starved for 48 h and preincubated for 30 min with different inhibitors, i.e. U0126 (20 nM), PP2 (15 μM), PP3 (15 μM), Ly29402 (50 μM), PD98059 (50 μM) and RU486 (10 μM). Afterwards 10 nM R5020 was added for 4 h (**b**). Cyclin D1 is induced non-genomically in a PR-dependent manner (**b**)

demonstrate that PRB interacts directly with the SH3 domain of src kinase. There was no interaction with the SH2 domain of src kinase (data not shown). The interaction with the SH3 domain of src alone, however, turned out to be hormone-independent. We only observed a hormone-dependent interaction of PRB with src kinase when we used a larger src construct that starts from the first amino acid of src and contains the SH3 as well as the SH2 domain (data not shown).

In order to screen for progestins with reduced non-genomic activity in vitro, we decided to use three different screening assays. To monitor genomic progestin effects, we used SKNMC cells that had been stably transfected with PRB and a luciferase reporter construct under the control of the mouse mammary tumour virus (MMTV) promoter. Dose–response curves were obtained for each progestin and the EC_{50} was determined.

To analyse non-genomic effects mediated by PRB, we used a mammalian two hybrid assay (MTH) exploiting the direct interaction of PRB and src kinase. We used PRB fused to the activation domain of NF-κB and src kinase fused to the DNA-binding domain of GAL4. The luciferase reporter gene was under the control of GAL4-responsive elements. The mammalian two hybrid experiments were performed in HeLa cells that were transiently transfected with the reporter gene plasmid and the two plasmids containing the fusion proteins. As already observed in our GST-pulldown experiments, the interaction of PRB with the SH3 domain of src kinase was hormone-independent in the MTH. When we used a longer src kinase construct, however, starting with amino acid one of src kinase and covering both the SH3 and SH2 domain, we observed a hormone-dependent interaction of PRB and src. We established the optimal assay conditions that allowed us to determine dose–response curves as well as EC_{50} values for each progestin.

As a second readout for non-genomic progestin action, we established a cyclin D1 luciferase assay using U2OS cells that were transiently transfected with PRB and a luciferase reporter gene under the control of the cyclin D1 promoter (generous gift of M. Beato). Interestingly, the induction of the cyclin D1-luciferase reporter construct by progestins was very sensitive to inhibition of PI3K and partly sensitive to src kinase and MAPK inhibition (data not shown). These results were in good agreement with the results obtained for the induction of cyclin D1 in T47D cells (Fig. 1b).

3 Identification and In Vitro Characterization of Tool Compounds

To identify progestins with reduced non-genomic activity in vitro, we screened a focussed library that contained 161 chemically diverse progestins. In all three in vitro assays, i.e. the transactivation assay (TA), the MTH assay and the cyclin D1 luciferase assay, we used the synthetic progestin R5020 as our reference compound. The EC_{50} values were determined and divided by the EC_{50} values of the reference progestin to yield the so-called competition factor (CF). By definition, the CF of R5020 in all three assays is 1. The primary criterion a compound had to fulfil in order to be subjected to further analysis was that it should stimulate the interaction of PRB/src with tenfold less potency than R5020, i.e. a CF of at least 10 should be achieved in the MTH as the primary criterion. To judge whether such a compound shows indeed a dissociation of genomic versus non-genomic effects in vitro, we divided the CFs for the MTH (monitoring non-genomic effects) by the CFs for the transactivation assay (monitoring genomic effects). Compounds with reduced genomic versus non-genomic effects should have a CF quotient larger than 1.

As depicted in Table 1, 4 out of 161 screened progestins stimulated the interaction of PRB with src with at least tenfold less potency than the reference compound R5020 and showed in addition a dissociation of their non-genomic (MTH) versus genomic effects (transactivation assay). The structures of the reference progestin R5020 and compounds A and B are depicted in Fig. 2. The structures of the non-steroidal compounds C and D are not yet disclosed. Compounds A, B and C showed a two- to fivefold better dissociation of their genomic versus non-genomic effects than R5020 as evidenced by their respective CF quotients (Table 1, column 4). The dissociation of genomic versus non-genomic effects of compound D was not as good as with the other compounds, but compound D was included for further analysis since it was by far the weakest activator in the second non-genomic assay, the cyclin D1 luciferase assay (Table 1, column 5). It should be noted that compound B, but none of the other compounds, showed strong glucocorticoid activity in vitro.

4   Testing of Tool Compounds In Vivo

To analyse how the reduced non-genomic activity in vitro translates into the in vivo situation, we used a mouse model that allowed for quantitative measurement of progestin action in the uterus and the mammary gland. Six-week-old C57BL/6 mice were ovariectomized. Starting 2 weeks post ovariectomy, mice were treated subcutaneously for 3 weeks with vehicle, 100 ng E2 or combinations of 100 ng E2 with different dosages of the respective progestin. Animals were injected with BrdU 2 h before sacrifice. As readout parameters for mammary

Table 1 Identification of tool compounds with reduced non-genomic activity in vitro

Compound	CF TA PRB	CF MTH PRB-src	$\frac{\text{CF MTH}}{\text{CF TA}}$	CF Cyclin D1 luciferase	Gluco-corticoid activity
R5020	1	1	1	1	no
A	15	71	4.4	4.7	no
B	6	29	4.8	3.9	yes
C	14	34	2.4	6.8	no
D	8	10	1.3	15.6	no

CF, EC_{50} compound divided by EC_{50} R5020

TA PRB, transactivation assay with progesterone receptor B (PRB)

MTH PRB-src, mammalian two hybrid assay based on interaction of PRB and src kinase

gland action we used ductal sidebranching and mammary epithelial cell proliferation, as well as changes in gene expression [i.e. repression of the E2-dependent gene indoleamine-pyrrole 2,3 dioxygenase (INDO) by progestins, and induction of cyclin D1]. Within the same animals we used the following readout parameters for uterine progestin action: (1) inhibition of E2-stimulated uterine epithelial cell proliferation and (2) inhibition of the expression of the E2-induced gene lactotransferrin (LTF). The following progestin doses were used in combination with 100 ng E2: medroxyprogesterone acetate (MPA) (0, 0.15, 0.75, 5, 10, 30 mg/kg), compound A (0, 2, 8, 20, 40, 120 mg/kg), compound B (0, 0.03, 0.2, 1, 2, 6 mg/kg), compound C (0, 0.2, 0.8, 5, 10, 25 mg/kg) and compound D (0, 0.2, 2, 10, 20, 60 mg/kg). Group size was eight animals per group. We used MPA as the reference progestin in the in vivo studies since it is widely used in hormone therapy and it was our aim to profile the newly identified progestins against MPA.

As a second independent readout parameter for uterine progestin action we performed maintenance of pregnancy assays in a second strain of mice.

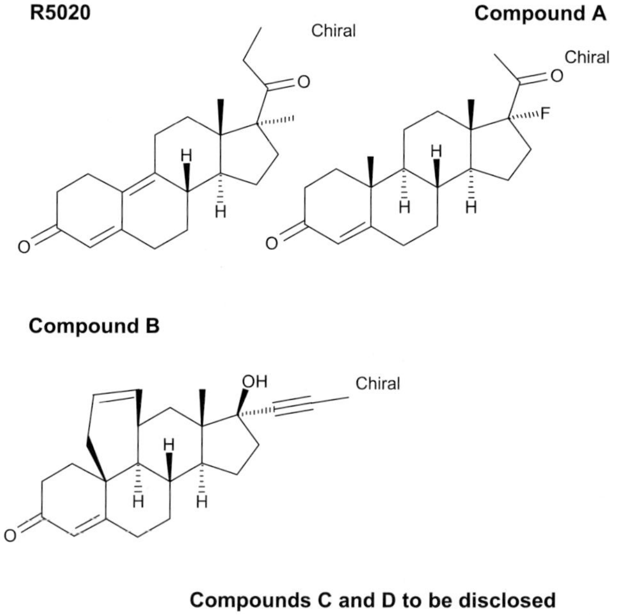

Fig. 2. Structures of the reference and tool compounds

4.1 Mammary Gland Assay

4.1.1 Ductal Sidebranching

Animals were sacrificed after 3 weeks of treatment with either vehicle, 100 ng E2 or 100 ng E2 plus different doses of each progestin. One inguinal mammary gland was processed for whole mount staining using carmine alum. Representative pictures of the mammary gland whole mount preparations are shown in Fig. 3. After vehicle treatment, only a few ducts were visible in the mammary fat pad, whereas after treatment with estradiol ducts started to elongate and end buds developed. Addition of increasing progestin doses led to further increase in ductal sidebranching. Tertiary and secondary sidebranches were quantified and dose–response curves were generated using Sigmaplot. The ED_{50} values for each progestin are depicted in Table 2. Unlike MPA, com-

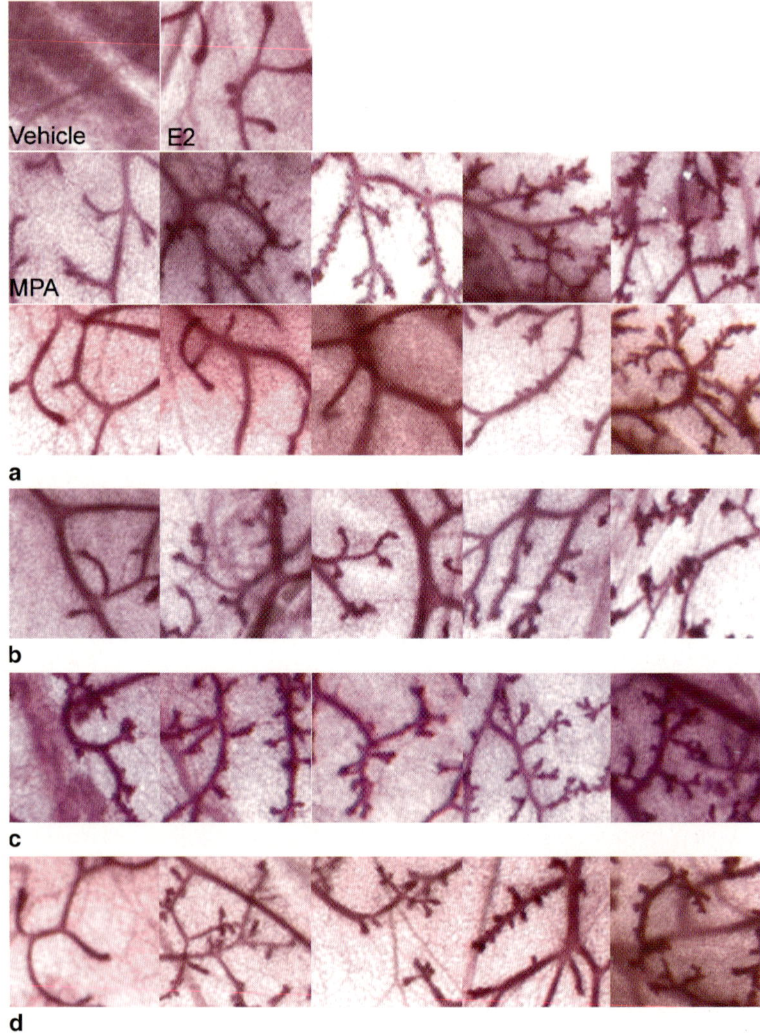
Vehicle
E2
MPA
a
b
c
d

Fig. 3. Analysis of ductal sidebranching. Six-week-old mice were ovariec-tomized. Starting 2 weeks post ovariectomy, animals were treated subcutaneously for 3 weeks with vehicle, 100 ng E2 or 100 ng E2 plus increasing doses of either MPA (0.15, 0.75, 5, 10, 30 mg/kg), compound A (2, 8, 20, 40, 120 mg/kg), compound B (0.03, 0.2, 1, 2, 6 mg/kg), compound C (0.2, 0.8, 5, 10, 25 mg/kg) or compound D (0.2, 2, 10, 20, 60 mg/kg). Increasing doses of each progestin are depicted in each row from the *left* to the *right*. Magnifications of representative whole mounts are shown

pounds C and D did not show 100% efficacy with regard to ductal side-branching. While compound C showed an efficacy of 75%, only 44% was reached with compound D (Table 2).

Table 2 Quantitative readouts for uterine and mammary gland effects of MPA and the tool compounds

Readout	MPA [mg/kg]	A [mg/kg]	B [mg/kg]	C [mg/kg]	D [mg/kg]
Uterus:					
Maintenance of pregnancy, ED_{50}	3.6	30	0.55	2.1	6
Epithelial cell proliferation, ID_{100}	2.9	14	0.12	0.5	6
Inhibition LTF expression, ID_{50}	0.7	1.58	0.11	0.45	0.1
Mammary gland:					
Ductal sidebranching, ED_{50}	3.4	50.4	1.8	0.36^a	1.28^a
Epithelial cell proliferation, ED_{50}	1.2	46.6	0.11	0.05^b	0.32^b
Inhibition INDO expression, ID_{50}	0.09	3.1	0.02	0.07	0.12
Cyclin D1 expression, first significant effect	0.75	20	No effect	No effect	No effect

[a] 75% efficacy for compound C, 44% efficacy for compound D
[b] 58% efficacy for compound C, 17% efficacy for compound D

4.1.2 Mammary Epithelial Cell Proliferation

To evaluate mammary epithelial cell proliferation, the dorsal two-thirds of the right inguinal mammary gland were fixed in 4% buffered formalin and embedded in paraffin. Sections were stained with an anti-BrdU antibody and proliferating cells in ducts were counted on four sections per animal. Dose–response curves were generated and ED_{50} values determined using Sigmaplot. The results are shown in Table 2. MPA as well as the compounds B, C and D had the tendency to stimulate the proliferation of mammary epithelial cells with an ED_{50} value that was smaller than the respective ED_{50} value for the stimulation of ductal sidebranching (Table 2). This phenomenon points to rather high relative proliferative activity in the mammary gland. However, compounds C and D—although stimulating mammary epithelial cell proliferation with high potency—did not reach 100% efficacy in this readout paradigm. Compound C reached only 58% efficacy, and compound D had only 17% efficacy if compared to MPA (Table 2).

4.1.3 Target Gene Induction in the Mammary Gland

For the analysis of gene expression, the ventral third of the right inguinal mammary (without lymph node) gland was rapidly frozen in liquid nitrogen. TaqMan-PCR analysis was performed and gene expression was normalized to cytokeratin-18 to correct for varying amounts of epithelial cells in the different samples. The results for the inhibition of E2-induced INDO expression in the mammary gland are shown in Table 2 whereas the results for cyclin D1 induction are shown in Table 2 and Fig. 4, respectively. Interestingly, compounds B, C and D did not show any additive effect on cyclin D1 compared to estradiol-only treatment. In contrast, MPA (showing the first significant effect at 0.75 mg/kg) and compound A (showing the first significant effect on cyclin D1 induction at 20 mg/kg) clearly induced cyclin D1 expression above those levels obtained after estradiol-only treatment (Fig. 4).

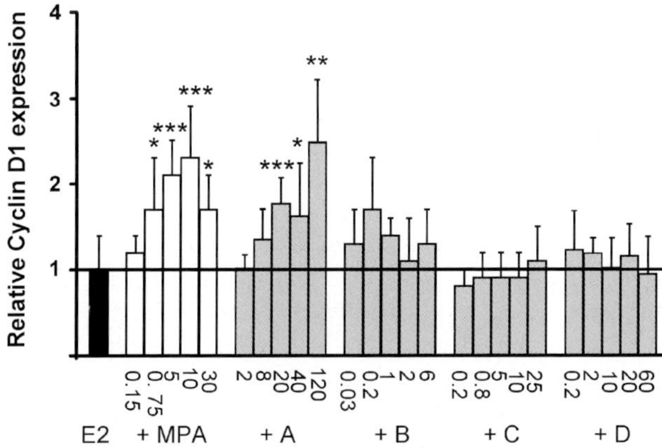

Fig. 4. Cyclin D1 induction in mammary glands. Cyclin D1 mRNA expression was analysed by TaqMan PCR using the ventral third of the right inguinal mammary gland from ovariectomized mice that had been treated for 3 weeks with either 100 ng E2 or 100 ng E2 plus increasing doses of different progestins. Cyclin D1 expression was normalized to cytokeratin-18 expression to correct for different epithelial cell content within the probes. Compound B, C and D did not induce cyclin D1 significantly above the levels seen with estradiol-only treatment. MPA at 0.75 mg/kg and compound A at 20 mg/kg showed a significant induction of cyclin D1 mRNA if compared to estradiol-only treatment. $*p<0.05$, $**p<0.01$, $***p<0.005$, two-sided Students t-test

4.1.4 Inhibition of Uterine Epithelial Cell Proliferation

To assess uterine progestin action in the same animals that had been subjected to the mammary gland whole mount assay, one uterine horn was fixed in 4% formalin and embedded in paraffin. Results of anti-BrdU stainings are shown in Fig. 5. After vehicle treatment, only single cells have reached the S-phase of the cell cycle (Fig. 5a), whereas after treatment with 100 ng estradiol for 3 weeks, approximately 25%–30% of the uterine epithelial cells proliferate. Increasing doses of progestins led to a reduction of uterine epithelial cell proliferation back to vehicle levels. Whereas treatment with 0.15 mg/kg MPA (Fig. 5c) did not inhibit uterine epithelial cell proliferation, there was a clear reduction in proliferation following 0.75 mg/kg (Fig. 5d), and with 5 mg/kg MPA

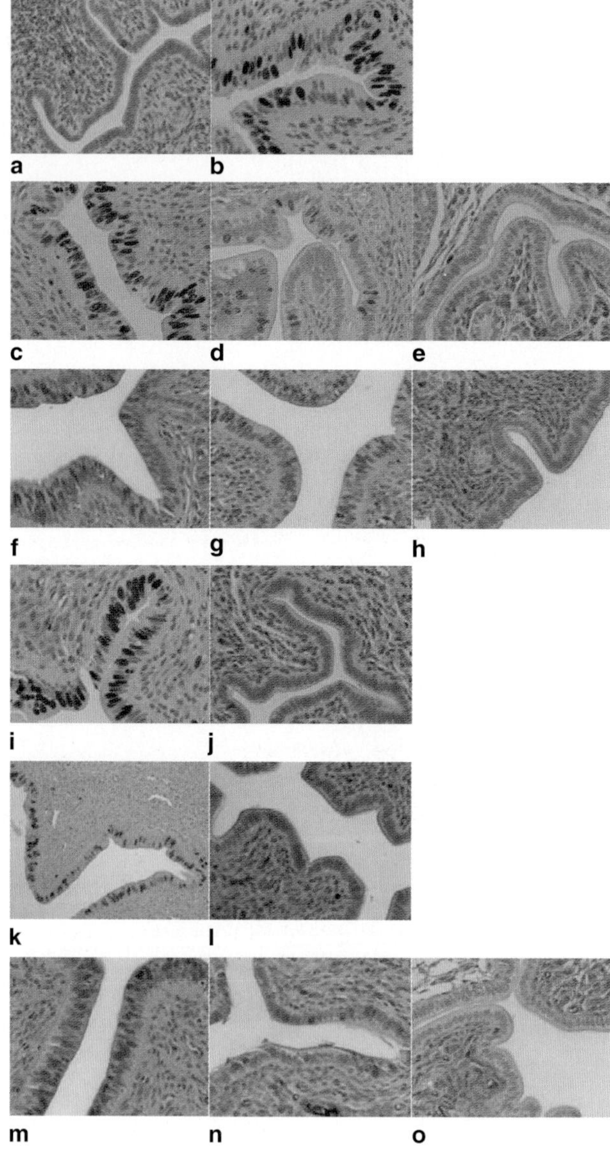

Fig. 5a–o. Inhibition of uterine epithelial cell proliferation by progestins. Uteri from mice sacrificed for the mammary gland assays were embedded in paraffin and stained with an antibody against BrdU. Uteri from animals treated with vehicle (**a**), 100 ng E2 (**b**) or 100 ng E2 plus increasing doses of progestins are shown, i.e. MPA 0.15 mg/kg (**c**), 0.75 mg/kg (**d**), 5 mg/kg (**e**), compound A 2 mg/kg (**f**), 8 mg/kg (**g**), 20 mg/kg (**h**), compound B 0.03 mg/kg (**i**), 0.2 mg/kg (**j**), compound C 0.2 mg/kg (**k**), 0.8 mg/kg (**l**), or compound D 0.2 mg/kg (**m**), 2 mg/kg (**n**), 10 mg/kg (**o**). Note that increasing amounts of progestins led to an inhibition of estradiol-induced uterine epithelial cell proliferation back to vehicle levels

cellular proliferation was suppressed back to vehicle levels (Fig. 5e). To calculate a mean ID_{100} for the inhibition of uterine epithelial cell proliferation, we calculated the mean of the dose-suppressing epithelial cell proliferation back to vehicle levels and the dose preceding this dose. In the case of MPA, the mean ID_{100} was 2.9 mg/kg. The ID_{100} values for the other progestins were calculated in analogy and are depicted in Table 2.

4.1.5 Inhibition of E2-Induced LTF Expression in the Uterus

To analyse the inhibition of E2-induced LTF expression in the uterus, the second uterine horn was rapidly frozen in liquid nitrogen. TaqMan-PCR was performed and dose–response curves were generated. RNA expression levels were normalized to the expression of TATA box binding protein. The ID_{50} values for each progestin are depicted in Table 2.

4.2 Maintenance of Pregnancy

To obtain a second independent readout for uterine progestin action, we performed maintenance of pregnancy assays. Adult female NMRI mice were mated with fertile males. The appearance of the vaginal plug was considered day 1 of pregnancy. On day 1 p.c. females were separated from the males and randomly assigned to different treatment groups (n=8 females per group). On day 8 of pregnancy the animals were ovariectomized and daily substituted subcutaneously with different doses of each progestin in combination with 0.03 µg estrone per animal beginning 2 h before ovariectomy. Ovariectomy without hormonal substitution would result in immediate abortion. On day 18 p.c. animals were sacrificed and living embryos were counted. The amount of living embryos per mouse in the sham-ovariectomized, vehicle-treated control group was set to 100%. Dose–response curves for MPA and drospirenone were generated using SigmaPlot. For each progestin, the ED_{50} values are shown in Table 2. Comparison of the ED_{50} values for the maintenance of pregnancy and the more semiquantitative ID_{100} values for inhibition of uterine epithelial cell proliferation showed that both sets of data were quite similar, although the experiments were performed in different strains of mice (Table 2). For the quantification of the dissociation of uterine versus mammary gland effects, we switched to the more quantitative ED_{50} values obtained from the maintenance of pregnancy assays.

4.3   Dissociation of Uterine Versus Mammary Gland Effects

To assess the dissociation between uterine and mammary gland effects for each compound in comparison to MPA, we calculated DFs by dividing ED_{50} values for mammary gland readouts by ED_{50} or ID_{50} values for uterine readouts (Table 2). By definition, a dissociation factor of 1 indicates that the respective progestin shows activity in the uterus and the mammary gland at the same dose. A dissociation factor smaller than 1 implies that the progestin is active in the mammary gland at lower doses than those required for activity in the uterus whereas dissociation factors larger than 1 indicate that the progestin exerts uterine effects at doses smaller than those required for considerable activity in the mammary gland. For hormone therapy, the latter profile (DF>1) would be highly desirable.

Whenever possible, the following rules were applied for the calculation of absolute DFs : (1) only readouts of the same type were compared, i.e. functional readouts were only compared with other functional readouts and molecular readouts were compared with other molecular readouts; (2) division of ED_{50} values by ED_{50} values and ID_{50} values by ID_{50} values. The calculated dissociation factors are depicted in Table 3.

For all different combinations of DFs, MPA showed values that were smaller than 1, indicating that MPA had the tendency to develop activity in the mammary gland at doses that are smaller than those required for uterine activity.

Compound A was the only compound that showed in almost all DFs values larger than 1. In comparison to MPA, compound A showed a better dissociation of uterine versus mammary gland effects. Compound A developed activity in the mammary gland at doses higher than those required for uterine activity and thus reflected the ideal profile of a progestin used for hormone therapy.

Compound B did not induce cyclin D1, but has values smaller than 1 in the DFs measuring proliferative activity in the mammary gland, i.e. ED_{50} BrdU mammary gland divided by ID_{100} BrdU uterus or divided by ED_{50} maintenance of pregnancy (Table 3). With regard to ductal sidebranching, compound B showed a better DF than MPA. In other words, compound B had relatively high proliferative activity in the mammary gland, higher than could be expected from its ability to stimulate ductal sidebranching. The relative high proliferative activity of compound B

Table 3 Calculation of dissociation factors (DFs) for mammary gland versus uterine progestin action

Calculation of DF[a]	MPA	A	B	C	D
$\dfrac{ED_{50}\,\text{sidebranching}}{ED_{50}\,\text{pregnancy}}$	0.98	1.7	3.3	0.17^{b}	0.21^{b}
$\dfrac{ED_{50}\,\text{BrdU Ma.}}{ID_{50}\,\text{BrdU ut.}}$	0.44	3.3	0.9	0.1^{b}	0.05^{b}
$\dfrac{ED_{50}\,\text{BrdU Ma.}}{ED_{50}\,\text{pregnancy}}$	0.33	1.6	0.2	0.02^{b}	0.05^{b}
$\dfrac{ID_{50}\,\text{INDO Ma.}}{ID_{50}\,\text{LTF ut.}}$	0.13	2.0	0.2	0.16	1.2
$\dfrac{\text{Cyclin D1 Ma.}}{ED_{50}\,\text{pregnancy}}$	0.21	0.7	—	—	—

[a] DFs are calculated by dividing quantitative measures of mammary gland action (*Ma.*) by those of uterine action (*ut.*); absolute DF values larger than 1 indicate favourable dissociation of uterine versus mammary gland effects
[b] Compounds show dissociation with regard to efficacy, not with regard to potency

in the mammary gland might be the result of its strong glucocorticoid activity in vitro. It has been demonstrated that mice lacking the glucocorticoid receptor in mammary epithelial cells showed an impaired proliferative response (Wintermantel et al. 2005). Therefore, glucocorticoid activity of compound B may lead to the enhanced proliferative activity in mammary epithelial cells.

The nonsteroidal compounds C and D showed a profile that is distinctly different. In contrast to all other compounds, the efficacies of compounds C and D with regard to stimulation of ductal sidebranching or with regard to stimulation of mammary epithelial cell proliferation never reached 100% (Table 3). Both compounds stimulated proliferative effects in the mammary gland with very high potency, the respective DFs were even smaller than those of MPA; however, there wass a remarkable efficacy dissociation, especially for compound D. Over a large dose range, compound D reached only 44% efficacy with regard to ductal sidebranching and 17% efficacy with regard to stimulation of mammary epithelial cell proliferation.

Taken together, we identified four progestins that showed reduced non-genomic versus genomic effects in vitro. These compounds were tested in vivo in comparison to MPA with regard to their dissociation of uterine versus mammary gland effects. Only one out of the four compounds, compound B, had an in vivo profile that was not better than the in vivo profile of MPA. Most likely, the strong glucocorticoid activity of compound B provoked its high proliferative activity in the mammary gland.

Among the other three other compounds, compound A showed a better dissociation of uterine versus mammary gland effects with regard to potency if compared to MPA, whereas compounds C and D showed a better dissociation of uterine versus mammary gland effects with regard to efficacy if compared to MPA.

5　　Outlook

Progestins with reduced proliferative activity in the mammary gland would be valuable tools for hormone therapy of postmenopausal women still having an uterus. Here we followed a screening hierarchy that was based on the hypothesis that reduced non-genomic activity of progestins may lead to reduced proliferative activity in the mammary gland. It has to be noted that this hypothesis might not be absolutely valid since in the normal mammary gland cell proliferation is induced in a paracrine manner and proliferating cells do not express the PR, as is the case in mammary cancer cells. Nevertheless, three out of four validated hits showed a dissociation of uterine versus mammary gland effects that was better than the dissociation seen with MPA. From our approach we have learned that progestins with glucocorticoid activity showed enhanced proliferation of mammary epithelial cells. Such progestins are not ideal for hormone therapy. The open question remaining is how might happen that a wrong hypothesis can drive a successful screen. We think that the decision to select only progestins for further analysis that stimulated the interaction of PRB with src with tenfold reduced potency if compared to R5020 had major impact on the outcome of the screen. We did not look for dissociation in a first step, but for tenfold reduced potency in comparison to R5020. In a second step we analysed whether the compounds showed a dissociation of genomic versus non-genomic effects. This approach led to a preselection of compounds that were less potent than R5020 and did not represent highly potent progestins. Most likely, progestins of intermediate strength, but not highly potent progestins, have a chance to show a dissociation of their activities in various tissues.

References

Anderson E, Clarke RB (2004) Steroid receptors and cell cycle in normal mammary epithelium. J Mammary Gland Biol Neoplasia 9:3–13

Bagowski CP, Myers JW, Ferrell JE Jr (2001) The classical progesterone receptor associates with p42 MAP kinase and is involved in phosphatidylinositol 3-kinase signaling in Xenopus oocytes. J Biol Chem 276:37708–37714

Boonyaratanakornkit V, McGowen E, Sherman L, Mancini MA, Cheskis BJ, Edwards DP (2007) The role of extranuclear signaling actions of progesterone receptor in mediating progesterone regulation of gene expression and the cell cycle. Mol Endocrinol 21:359–375

Boonyaratanakornkit V, Scott MP, Ribon V, Sherman L, Anderson SM, Maller JL, Miller WT, Edwards DP (2001) Progesterone receptor contains a praline rich motif that directly interacts with SH3 domains and activates c-src family tyrosine kinases. Mol Cell 8:269–280

Castoria G, Barone MV, Di Domenico M, Bilancio A, Ametrano D, Migliaccio A, Auricchio F (1999) Non-transcriptional action of oestradiol and progestin triggers DNA synthesis. EMBO J 18:2500–2510

Falkenstein E, Tillmann HC, Christ M, Feuring M, Wehling M (2000) Multiple actions of steroid hormones—a focus on rapid, nongenomic effects. Pharmacol Rev 52:513–556

Hofseth LJ, Raafat AM, Osuch JR, Pathak DR, Slomski CA, Haslam SZ (1999) Hormone replacement therapy with estrogen or estrogen plus medroxyprogesterone acetate is associated with increased epithelial proliferation in the normal postmenopausal breast. J Clin Endocrinol Metab 84:4559–4565

Hulka BS, Chambless LE, Kaufman DG, Fowler WC Jr, Greenberg BG (1982) Protection against endometrial carcinoma by combination-product oral contraceptives. JAMA 247:475–477

Migliaccio A, Piccolo D, Castoria G, Di Domenico M, Bilancio A, Lombardi M, Gong W, Beato M, Auricchio F (1998a) Activation of the src/p21ras/ERK pathway by progesterone receptor via cross-talk with estrogen receptor. EMBO J 17:2008–2018

Migliaccio A, Piccolo D, Castoria G, Di Domenico M, Bilancio A, Lombardi M, Gong W, Beato M, Auricchio F (1998b) Activation of the src/p21ras/ERK pathway by progesterone receptor via cross-talk with estrogen receptor. EMBO J 17:2008–2018

Mulac-Jericevic B, Lydon JP, DeMayo FJ, Conneely OM (2003) Defective mammary gland morphogenesis in mice lacking the progesterone receptor B isoform. Proc Natl Acad Sci USA 100:9744–9749

Said TK, Conneely O, Medina D, O'Malley BW, Lydon JP (1997) Progesterone, in addition to estrogen, induces cyclin D1 expression in the murine mammary epithelial cell, in vivo. Endocrinology 138:3933–3939

Wintermantel TM, Bock D, Fleig V, Greiner EF, Schütz G (2005) The epithelial glucocorticoid receptor is required for the normal timing of cell proliferation during mammary lobuloalveolar development but is dispensable for milk production. Mol Endocrinol 19:340–349

Ernst Schering Foundation Symposium Proceedings, Vol. 1, pp. 171–197
DOI 10.1007/2789_2008_078
© Springer-Verlag Berlin Heidelberg
Published Online: 29 February 2008

In Vitro and In Vivo Characterization of a Novel Nonsteroidal, Species-Specific Progesterone Receptor Modulator, PRA-910

Z. Zhang, S.G. Lundeen, O. Slayden, Y. Zhu,
J. Cohen, T.J. Berrodin, J. Bretz, S. Chippari, J. Wrobel,
P. Zhang, A. Fensome, R.C. Winneker, M.R. Yudt[✉]

Women's Health and Musculoskeletal Biology, Wyeth Research, 500 Arcola Road, 19426 Collegeville, USA
email: *yudtm@wyeth.com*

Abstract. The progesterone receptor (PR) is an important regulator of female reproduction. Consequently, PR modulators have found numerous pharmaceutical utilities in women's reproductive health. In the process of identifying more receptor-specific and tissue-selective PR modulators, we discovered a novel nonsteroidal, 6-aryl benzoxazinone compound, PRA-910, that displays unique in vitro and in vivo activities. In a PR/PRE reporter assay in COS-7 cells, PRA-910 shows potent PR antagonist activity with an IC_{50} value of approximately 20 nM. In the alkaline phosphatase assay in the human breast cancer cell line T47D, PRA-910 is a partial progesterone antagonist at low concentrations and is also an effective PR agonist at higher concentrations (EC_{50} value of approximately 700 nM). PRA-910 binds to the human PR with high affinity (K_d=4 nM) and was previously shown to exhibit greater than 100-fold selectivity for the PR versus other steroid receptors. In the adult ovariectomized rat, PRA-910 is a potent PR antagonist. It inhibits progesterone-induced uterine decidual response with an ED_{50} value of 0.4 mg/kg, p.o., and reverses progesterone suppression of estradiol-induced complement C3 expression with potency similar to RU-486. In the nonhuman primate, however, PRA-910 is a PR agonist. The effect on endometrial histology strongly resembles that of progesterone. This unique compound also suppresses estradiol-induced epithelial cell proliferation and both estrogen and progesterone receptor expression in the uterine endometrium as a PR agonist would. In summary, PRA-910 is a structurally and biologically novel selective PR modulator with either PR agonist or antagonist activity, depending on context, concentration, and species.

1   Introduction

Progesterone (P_4), acting via the progesterone receptor (PR), plays a pivotal role in female reproduction. It is involved in the regulation of uterine development and differentiation, ovum implantation, ovulation, and mammary gland development (Psychoyos 1973; Zhang et al. 1994; Lydon et al. 1995). Progesterone receptor agonists, i.e., progestins, have several therapeutic applications in women's health. Progestins are a major component of oral contraceptives, achieving efficacy through inhibition of ovulation and thickening of cervical mucus. In a post-menopausal hormone therapy paradigm, a progestin is primarily used in combination with an estrogen to protect the endometrium from estrogen-induced hyperplasia. PR antagonists, on the other hand, currently have only limited therapeutic applications. The only clinically approved PR antagonist, i.e., mifepristone or RU-486, is indicated for termination of early pregnancy in combination with a prostaglandin. A growing body of evidence, however, suggests that PR antagonists may have significant therapeutic potential. For example, preclinical evidence in primates suggests that PR antagonists may have utility in both oral contraception and postmenopausal hormone therapy (Slayden et al. 2001a, 2006). In addition, data from small clinical studies indicate that PR antagonists may be effective in the treatment of reproductive disorders such as uterine fibroids and endometriosis, as well as hormone-sensitive tumors (Chwalisz et al. 2005).

Another concern with the currently available PR antagonist, RU-486, is receptor selectivity (Beck et al. 1993). This compound has potent anti-glucocorticoid activity as well. Likewise, other steroidal PR modulators either commercially available or still in development face the same selectivity hurdle, though sometimes cross-reactivity with other steroid receptors, i.e., anti-androgenic or anti-mineralocorticoid activity, can be marketed as a positive attribute (Foidart 2000). Nevertheless, the side effects associated with steroid receptor cross-reactivity can limit long-term use and approvable indications for PR modulators. To this end, our drug discovery approach was to begin with nonsteroidal compounds in order to avoid potential receptor-selectivity issues and the common metabolic pathways steroidal compounds employ.

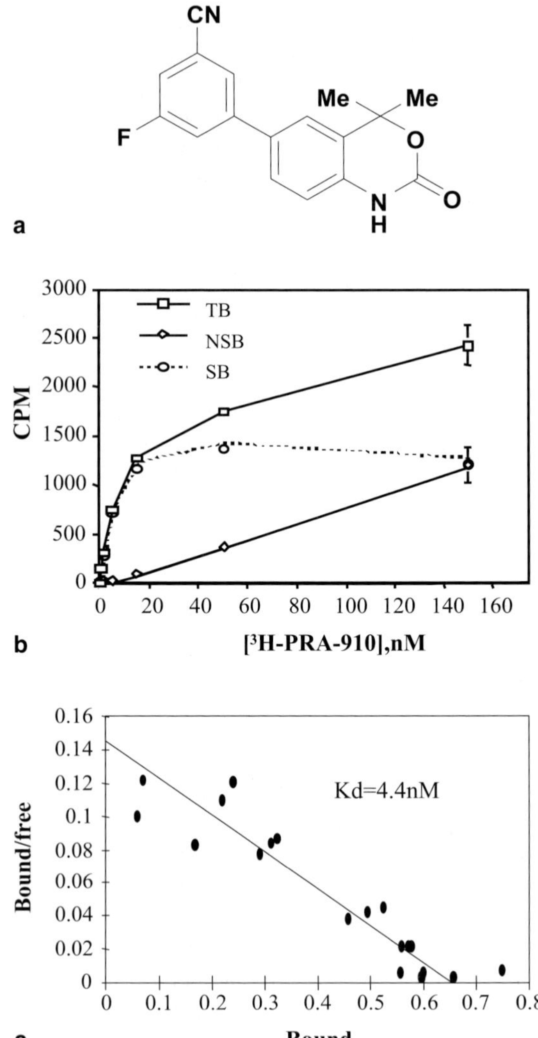

CN
Me   Me
F
O
N
H
O
a
3000
TB
NSB
SB
CPM
2500
2000
1500
1000
500
0
0   20   40   60   80   100 120 140 160
[³H-PRA-910],nM
b
0.16
0.14
0.12
0.1
0.08
0.06
0.04
0.02
0
Kd=4.4nM
Bound/free
0   0.1 0.2 0.3 0.4 0.5 0.6 0.7 0.8
Bound
c

Fig. 1 a. Structure of PRA-910. **b** Binding of PRA-910 to the human PR. Increasing concentrations of $[^3H]$-PRA-910 were incubated with T47D cell cytosol in the presence or absence of 100-fold excess of unlabeled compound. Following separation of unbound compound, specific bound (*SB*) counts (*CPM*) were calculated by subtracting nonspecific binding (*NSB*) from total bound (*TB*) counts. **c** Scatchard analysis of $[^3H]$-PRA-910 binding to the human PR. The affinity constant of 4.4 nM calculated in this experiment is similar to that determined in two additional experiments

In this article, we present the in vitro and in vivo characterization of a novel nonsteroidal PR modulator, PRA-910 (Fig. 1a) that exhibits species- and context-specific activities. This unique molecule elicits potent PR antagonist activity in the rat but is a PR agonist in the nonhuman primate. In vitro, the mixed antagonist and agonist properties of PRA-910 depend on the assay context and concentration used. These data with PRA-910 highlight the complexity of progesterone signaling, provide evidence for species selectivity, and suggest a new molecular mechanism from which to modulate PR function.

2   Materials and Methods

2.1   Reagents

P_4, dexamethasone (DEX), flutamide, 17α-ethinyl estradiol (EE), and testosterone propionate (TP) were purchased from Sigma Chemical Co. (St. Louis, MO). Mifepristone (RU-486) was purchased from the Shanghai Organic Chemical Institute (China). Tissue culture media Dulbecco's modified Eagle's medium (DMEM)/F12, MEM, and DMEM were obtained from Invitrogen (Grand Island, NY). Fetal bovine serum (FBS) was obtained from Hyclone (Logan, UT). $[^3H]$-R5020 was purchased from NEN Life Science Products (Boston, MA). PRA-910 and $[^3H]$-PRA-910 were prepared by the Medicinal Chemistry and Pharmaceutical Development Departments, respectively, at Wyeth. All other chemicals were purchased from Sigma. Test compounds were dissolved in dimethylsulfoxide (DMSO) or a 50/50 (v/v) mixture of DMSO and ethanol vehicle for subsequent in vitro and in vivo studies.

2.2   Cell Culture and In Vitro Assays

Human breast carcinoma cell lines T47D and monkey kidney cell line COS-7, were obtained from American Type Culture Collection (Rockville, MD). T47D cells were maintained in DMEM/F12 and the COS-7 were maintained in DMEM containing 10% FBS, 0.1% MEM nonessential amino acids, 100 U/ml penicillin, 100 mg/ml streptomycin, and 2 mM GlutaMax (Invitrogen). Cells were passed 2–3 times every week.

2.3   PR Binding Assay

T47D cells were homogenized in 20 mM hydroxyethylpiperazine ethanesulfonic acid (HEPES) buffer (1 mM EDTA, pH 7.6) with protease inhibitors (aprotinin 0.5 μg/ml, leupeptin 0.5 μg/ml, pepstatin A 0.7 μg/ml, and PMSF 0.5 mM). Uterine tissues from rat, rabbit, and monkey were homogenized in 10 mM Tris buffer (1.5 mM EDTA, pH 7.4) with sodium molybdate and 1 mM dithiothreitol. After centrifugation at 100,000×g for 1 h (4°C), the supernatant containing PR was collected and measured for protein concentration. The PR binding assay was performed with 100 μg cytosolic protein with an increasing concentration of [^{3}H]-PRA-910 with or without unlabeled R5020 as a cold competitor. Following overnight incubation at 4°C, free and bound [^{3}H]-PRA-910 were separated using 1% charcoal/0.05% dextran 69K in Tris/EDTA buffer (pH 7.4). Total and bound [^{3}H]-PRA-910 were counted in a Beckman LS6500 Scintillation Counter (Beckman Instruments, Fullerton, CA).

2.4   PRE-Luciferase Assay

COS-7 cells (30,000 cells/well) were transfected overnight using FuGene (Roche Applied Science, Indianapolis, IN) with 0.1 μg/well of full-length human PR-B and 0.2 μg/well progesterone response element (PRE)-luciferase vector in 96 well plates (Zhang et al. 2002a). Cells were treated with ligands for 24 h and luciferase activity was measured on a Victor2 luminometer (Millipore, Billerica, MA) using the Luciferase Reporter assay kit (Promega, Madison, WI).

2.5   T47D Alkaline Phosphatase Assay

The effect of PRA-910 on the alkaline phosphatase activity in T47D
cells was determined as described previously (Zhang et al. 2000, 2002a).
Briefly, cells were plated in 96-well plates at 50,000 cells/well in
DMEM/F12 with 10% FBS. After overnight culture, the medium was
changed to phenol red-free DMEM/F12 containing 2% charcoal-
stripped FBS (experimental medium). The next day, cells were treated
with test compound in the presence (antagonist mode) or absence
(agonist mode) of 1 nM P_4 in the experimental medium. Cellular alkaline phosphatase activity was measured 24 h after treatment using *p*-
nitrophenyl phosphate as substrate. Optical density measurements were
taken at 5-min intervals for 30 min at a test wavelength of 405 nM.

2.6   PR/SRC-3 Mammalian Two Hybrid Assay

COS-7 cells were transfected with PR ligand binding domain (LBD) in
the GAL4 DNA binding domain plasmid pM (Clontech, Palo Alto, CA),
full-length steroid receptor coactivator (SRC)-3 in the VP16 activation
domain plasmid pVP16 (Clontech), and a GAL4 responsive luciferase
reporter (5×GALuas) using Lipofectamine 2000 (Invitrogen) according
to the manufacturer's instructions. Cells were treated with ligands for
24 h and luciferase activity was measured on a Victor2 luminometer
(Perkin Elmer Life Sciences) using the Luciferase Reporter assay kit
(Promega).

2.7   Protease Digestion Assay

The protease digestion analysis was performed essentially as described
(Allan et al. 1992; McDonnell et al. 1995) with minor modifications.
The plasmid pT7BPRB, kindly provided by Dr. O'Malley (Baylor College of Medicine), was used to generate [^{35}S]-radiolabeled PR-B using
the TNT T7 Quick Coupled Transcription/Translation System according to the manufacturer's protocol (Promega). After the translation reaction, an aliquot (30 µl) of the lysate was incubated for 10 min at room
temperature (RT) in the absence or presence of ligands at a final concentration of 100 nM. Aliquots (5 µl) of the ligand-treated receptor mixture

were then incubated with a trypsin solution (Worthington Biochemicals, Freehold, NJ), giving various final concentrations of the enzyme (0, 25, 50, 75 µg/ml). After incubation at RT for 10 min, the digestion reaction was terminated with the addition of 20 µl gel denaturing buffer and boiling for 5 min. The digestion products were separated on a 4–12% Bis–Tris NuPAGE gel (BioRad Laboratories, Hercules, CA). After the electrophoresis the gel was treated with a 50% (v/v) methanol–10% acetic acid (v/v) solution for 30 min and immersed in Amplify (Amersham Pharmacia Biotech, Piscataway, NJ) for 30 min. The gel was then dried under vacuum and the radiolabeled products were visualized by autoradiography.

2.8 In Vivo Studies

All animal studies were conducted under Wyeth or Oregon Regional Primate Research Center institutional approval protocols.

2.9 Rat Uterine Decidual Assay

Rat decidualization assay was run as described previously (Lundeen et al. 2001). In brief, mature female Sprague–Dawley rats (~220 g) were ovariectomized at least 10 days before treatment to reduce circulating sex steroids. PRA-910 was administered once daily for 7 days orally by gavage (0.5 ml) in 2% Tween 80/0.5% methyl-cellulose vehicle with concurrent administration of P_4 (5.6 mg/kg, s.c.). Approximately 24 h after the third daily treatment, decidualization was induced in one uterine horn of each anesthetized rat by scratching the antimesometrial luminal epithelium with a blunt 21-gauge needle. The contralateral horn was not scratched and served as a nonstimulated control. Animals were euthanized by CO_2 asphyxiation 24 h following the final treatment. The uteri were removed and trimmed of fat and the decidualized (D) and control (C) horns were weighed separately. The decidual response is expressed as D/C.

2.10   Rat Uterine Complement C3 Assay

Ovariectomized female 60-day-old Sprague–Dawley rats were obtained from Harlan (Indianapolis, IN). Ovariectomies were performed by the supplier a minimum of 8 days before treatment. The rats were randomized and placed in groups of 6. The animals were treated with PRA-910 once daily for 2 days orally by gavage in a volume of 0.5 ml with concurrent treatment of progesterone (3 mg/kg, s.c.). On the second day of treatment the animals were also treated with EE (0.08 mg/kg) orally by gavage. Approximately 24 h after the final treatment the animals were euthanized by CO_2 asphyxiation. The uteri were then removed, stripped of remaining fat and mesentery, weighed, and snap-frozen on dry ice. Total RNA was isolated from the uteri using the Trizol Reagent (GibcoBRL, Gaithersberg, MD) according to the manufacturer's instruction, and complement C3 expression was analyzed by Northern analysis as described (Lundeen et al. 2001).

2.11   Rhesus Monkey Study

A preliminary experiment was conducted to determine if PRA-910 acts as a PR antagonist to induce menses in artificial cycling rhesus macaques. Two ovariectomized macaques were implanted with a 5-cm silastic capsule packed with crystalline E_2 to stimulate an artificial follicular phase (Slayden et al. 1995). After 14 days of E_2 priming, a similar 6-cm P_4-filled implant was inserted to stimulate an artificial luteal phase. After 14 days of sequential E_2+P_4 priming, the animals were treated with 5 mg/kg PRA-910, i.m., for 10 days. A daily vaginal swab was done to check for evidence of menstruation for 18 days.

In order to determine if PRA-910 is a PR agonist in rhesus monkeys, two monkeys were treated with the compound (5 mg/kg, i.m.) for 7 days after 14 days of E_2 implant. At the end of treatment each animal received 100 mg Br(d)U i.v. Following treatments, rhesus monkeys were euthanized and reproductive tracts were collected (Slayden et al. 2006). Tissue samples for morphological analyses were prepared as previously described (Slayden et al. 1995, 2001a). In brief, the microwave-treated sections were lightly fixed (0.2% picric acid, 2% paraformaldehyde in PBS) for 10 min; the immunocytochemistry (ICC) was conducted with monoclonal anti-ER (1D-5; BioGenex, San Ramon, CA), anti-PR (anti-PR Ab-8; NeoMarkers; Fremont, CA), anti-Ki-67 antigen (Dako, Carpinteria, CA) or anti-Br(d)U (ICN Biomedical Research Products, Costa Mesa, CA). In each case primary antibody was reacted with either biotinylated anti-mouseG second antibody and detected with an avidin-biotin peroxidase kit (Vector Laboratories, Burlingame, CA).

Photomicrographic images were captured through an Optronics DE 75 0T digital camera (Optronics Engineering, Goleta, CA) and digitized images were printed on an Epson Stylus Photo 1200 printer. Endometrial stromal compaction was quantified by counting stroma cells within 100×100 μm fields in the functionalis zone of each sample. Two methods were used to estimate the abundance of proliferating endometrial epithelial cells. First, the abundance of mitotic cells in glycol methacrylate (GMA) sections were counted in the functionalis and basalis zones. Second, the number of Br(d)U-labeled cells were counted from samples collected at necropsy. ICC sections stained for Ki-67 antigen were also prepared and photographed, but because Ki-67 staining appeared to parallel mitotic counts, counts of Ki-67-positive cells were not made. Due to the small sample size for each group, no statistical analysis was conducted and all numeric values are reported as means. However, these mean values were compared to the naturally occurring variance we have previously observed in values from control animals from other studies (Slayden et al. 2006).

2.12   Rat Glucocorticoid Receptor Activity Assay

Male Sprague–Dawley rats, adrenalectomized by the vendor (Taconic Farms, NY), were shipped 2 days after surgery and acclimated for 2 days before beginning the study. Average body weights were 140 to 160 g upon arrival. The animals were maintained on a diet of standard rodent laboratory chow and 0.85% saline drinking water. Animals were weighed and grouped (6 per group). PRA-910 alone or concurrent with dexamethasone (s.c., 0.05 mg/kg) was administered orally in 2% Tween 80/0.5% methyl cellulose vehicle in a volume of 0.5 ml/rat. Animals were treated once daily for five consecutive days. Animals were fasted 16 h before necropsy. Approximately 7 h after final dosing, the animals were euthanized by CO_2 asphyxiation. The thymus was removed and its weight recorded.

2.13 Rat Androgen Receptor Activity Assay

Immature male Sprague–Dawley rats, castrated by the vendor (Taconic Farms, NY) (~22 days old, 40–45 g body weight), were shipped 2 days after surgery and acclimated for 5 days before beginning the study. The animals were maintained on a diet of standard lab chow and water. The animals were weighed and grouped 7 days after castration, 5–6 animals per group, to provide similar mean body weights. To assess its anti-androgenic activity, PRA-910 was administered once daily for 10 consecutive days by gavage in a volume of 0.5 ml per rat in the 2% Tween 80/0.5% methylcellulose with a concurrent dose (1.0 mg/kg s.c.) of the reference androgen testosterone propionate. Flutamide (FLU) was administered s.c. at 10 mg/kg as a positive anti-androgenic control. Approximately 24 h after the final dosing, animals were sacrificed by CO_2 asphyxiation. The ventral prostate and seminal vesicles were excised, cleaned of extraneous tissue, blotted on filter paper, and weighed.

2.14 Evaluation of Results

In T47D alkaline phosphatase assay, a dose–response curve was generated for dose (x-axis) vs the rate of enzyme reaction (slope) (y-axis) for test compounds. Square root-transformed data were used for analysis of variance and nonlinear dose–response curve fitting in the PR binding assay, T47D cell alkaline phosphatase and the PR-B/PRE cotransfection assay. Huber weighting was used to down-weight the effects of outliers. EC_{50} or IC_{50} values were calculated from the re-transformed values. JMP software (SAS Institute, Cary, NC) was used for both one-way analysis of variance and nonlinear dose–response analysis. Data for the rat uterine decidualization, glucocorticoid, and androgenic assays were transformed by logarithms to maximize normality and homogeneity of variance. The Huber M-estimator was used to down-weight the outlying transformed observations for both dose–response curve fitting and one-way analysis of variance (ANOVA) using the JMP software (SAS

Institute). Values of ED_{50} (50% effective dose) were calculated from the transformed values.

3 Results

3.1 PRA-910 Binds to the Progesterone Receptor from Multiple Species

The synthesis and initial description of PRA-910 (Fig. 1a) activity has been previously described (Zhang et al. 2002b). For quantitative affinity measurements, [^{3}H]-PRA-910 was synthesized and used in binding studies with T47D human breast cancer cell cytosol as the source of human PR. Saturation binding and Scatchard plot analysis shows that PRA-910 has saturable, specific binding sites in T47D cell cytosolic preparations with a K_d value of 4.4 nM (Fig. 1b and c). In a PR competition binding assays using cytosol prepared from monkey, rat, or rabbit uterus, PRA-910 inhibited [^{3}H]-R5020 binding dose dependently with IC_{50} values of 23, 14, and 161 nM, respectively (data not shown).

3.2 Functional Activity of PRA-910 on PR in COS-7 and T47D Cells

The in vitro biological activity of PRA-910 was evaluated in two cell-based assays, a PR/PRE cotransfection assay in COS-7 cells and the T47D alkaline phosphatase assay. The compound was tested for both PR agonism and antagonism in these assays. In the COS-7 cells cotransfected with hPR-B and a PRE-luciferase reporter, PRA-910 showed potent PR antagonist activity with an IC_{50} (50% inhibition of 1 nM P_4-induced PRE-luciferase activity) value of approximately 20.6 nM but only 53% inhibition relative to the control steroidal antagonist RU-486 (Fig. 2a). In the same assay run in the agonist mode (compound alone), PRA-910 exhibited partial agonist activity at 300 and 1,000 nM (Fig. 2b).

In the T47D cells, which express both human PR-A and PR-B, measurement of alkaline phosphatase induction is an endogenous endpoint used to measure activity of PR modulators. Evaluation of PRA-910 in the T47D alkaline phosphatase assay resulted in a biphasic curve

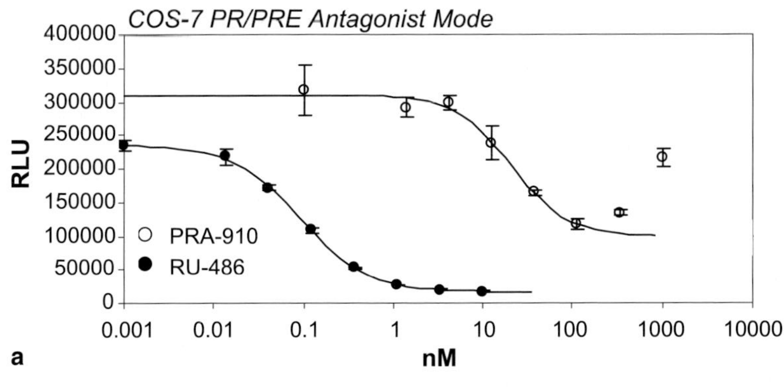

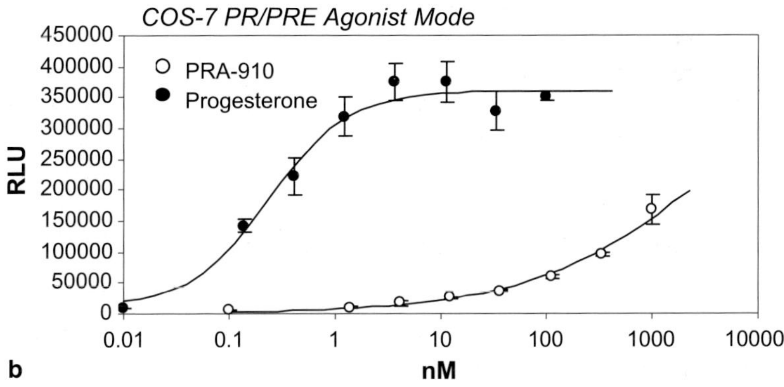

Fig. 2a–b. In vitro functional analysis of PRA-910. **a** COS-7 cells were cotransfected with PR-B and PRE-luciferase reporter vectors and treated for 16 h with the indicated concentrations of PRA-910 (*open circles*) or RU-486 (*closed circles*) in the presence of 1 nM P_4. The IC_{50} values are 0.093+/–0.008 for RU-486 and 20.6+/–8.8 for PRA-910. **b** The PR-B/PRE cotransfection assay was run in agonist mode with either P_4 (*closed circles*) or PRA-910 (*open circles*) alone. The EC_{50} value for P_4 is 0.24+/–0.06. The curve for PRA-910 failed to plateau and the IC_{50} cannot be accurately estimated

(Fig. 2c). At low doses (<100 nM), PRA-910 antagonized the progesterone-induced rise in alkaline phosphatase activity with an average efficacy of 43% and an average IC_{50} of 14.3+/–8.1 nM (*n*=3). At concen-

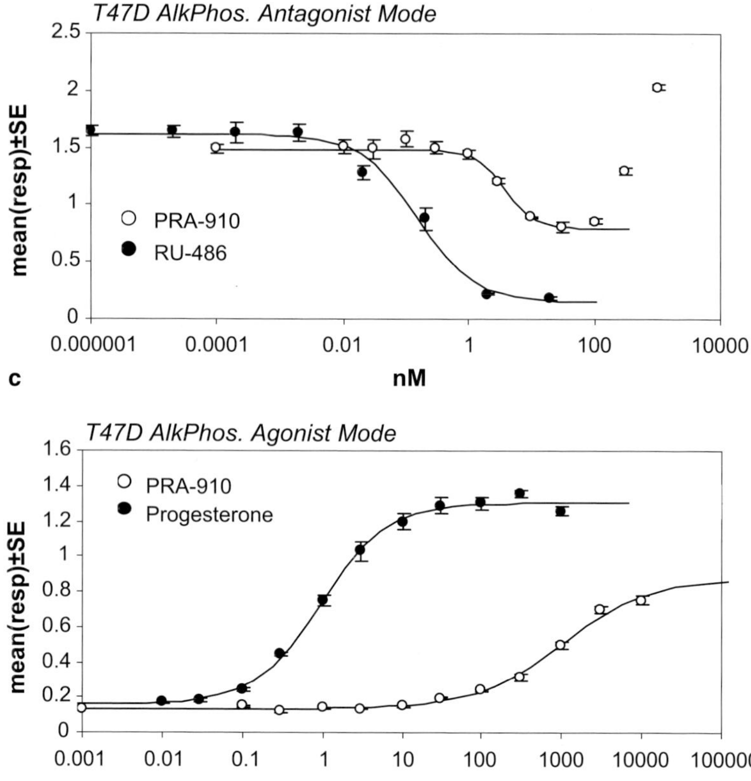

Fig. 2c–d. c T47D alkaline phosphatase assay in antagonist mode. Cells were treated with PRA-910 (*open circles*) or RU-486 (*closed circles*) in the presence of 1 nM P_4 for 16 h. The IC_{50} for RU-486 is 0.051+/–0.004 nM. The IC_{50} for PRA-910 is 18.9+/–2.9 nM. **d** T47D alkaline phosphatase assay in agonist mode. T47D cells were treated with P_4 (*closed circles*) or PRA-910 (*open circles*) alone. The EC_{50} for P_4 is 0.95+/–0.07 nM and the EC_{50} for PRA-910 is approx. 1.0 μM+/–0.2 μM, but failed to completely plateau at 10 μM. Each plot (**a–d**) is representative of at least three separate experiments

trations greater than 100 nM, PRA-910 acted as a partial PR agonist in alkaline phosphatase induction with an average EC_{50} of 762+/–247 nM (*n*=3) and approximately 60–70% efficacy of P_4 (Fig. 2d).

3.3 PRA-910 Promotes SRC-3 Coactivator Interaction with the PR

The peculiar biphasic shape of the antagonist dose–response curves in Fig. 2 prompted further investigation into the mechanism of action of PRA-910 on the PR. A mammalian two-hybrid assay was employed to measure the interaction between PR and the coactivator SRC-3. Consistent with the transcriptional readouts above, PRA-910 showed both PR agonist and antagonist activity in a dose-dependent fashion. As shown in Fig. 3, at lower concentrations (<100 nM), PRA-910 suppressed P_4-induced PR/SRC-3 interaction with an IC_{50} value of 21 nM and about 70% efficacy as compared to RU-486 (not shown). Under the same conditions, but in the absence of P_4, PRA-910 showed PR agonism as measured by recruitment of SRC-3 with an EC_{50} value of approx. 1,000 nM and approximately 75% the efficacy of P_4. In this assay both the agonist and antagonist plots overlap at concentrations above 100 nM.

3.4 Limited Proteolytic Analysis of PRA-910 Bound PR

Limited protease digestion was carried out to compare conformation changes of PR upon PRA-910 binding relative to the steroidal compounds P_4 and RU-486. As shown in Fig. 4, PRA-910 bound PR provided a trypsin digestion pattern that was similar to P_4 (agonist)-bound PR but distinct from that of RU-486 (antagonist) bound PR. The concentration of PRA-910 used is 100 nM, a concentration where full antagonist activity is observed in the cell-based assays. Higher concentrations of PRA-910 showed a similar pattern. Similar patterns were also observed when different pre-incubation times, i.e., overnight at 4°C, were done (not shown). Partial digestion with chymotrypsin also showed similar patterns for both PRA-910 and P_4-bound PR that are distinct from that of RU-486-bound PR (data not shown). The less intensely protected bands in PRA-910-treated lanes compared to P_4 probably reflect the slightly weaker binding affinity of PRA-910 or differences in kinetic constants (off rates) between the two molecules.

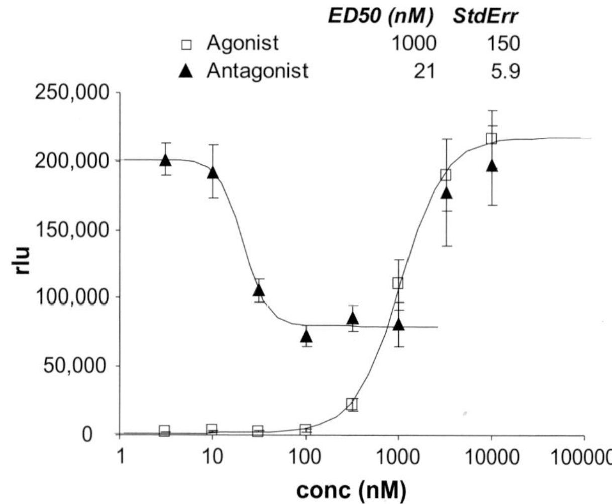

Fig. 3. A mammalian two-hybrid assay measuring the ligand-dependent interaction between the PR-LBD and SRC-3-VP16 coactivator. PRA-910 run in agonist mode (*open squares*) is a fully efficacious agonist in this assay with and ED_{50} of approximately 1,000 nM. When run in the antagonist mode (*closed triangle*) against 5 nM of the synthetic progestin R5020, PRA-910 achieves 70% inhibition with an ED_{50} of 21 nM, and changes profile to recruit coactivator at higher concentrations, overlapping with the agonist mode curve at concentrations above 300 nM. This is a representative plot of three separate experiments. The reference compounds not shown are P_4 in agonist mode with an $ED_{50}=3.7$ nM, and RU-486 in antagonist mode, $ED_{50}=0.3$ nM

3.5   PR Antagonist Activity of PRA-910 in the Rat

Two rat models were used to evaluate the PR activity of PRA-910. In the rat decidualization assay, the animals were treated with PRA-910 orally for 7 days with or without concurrent subcutaneous (s.c.) administration of P_4. PRA-910 suppressed the P_4-induced decidual response dose dependently with a mean ED_{50} (50% inhibition of progesterone induced decidual response) of 0.3+/–0.02 mg/kg ($n=5$), similar to the value for RU-486 (0.2 mg/kg; Fig. 5). PRA-910, when dosed alone, did

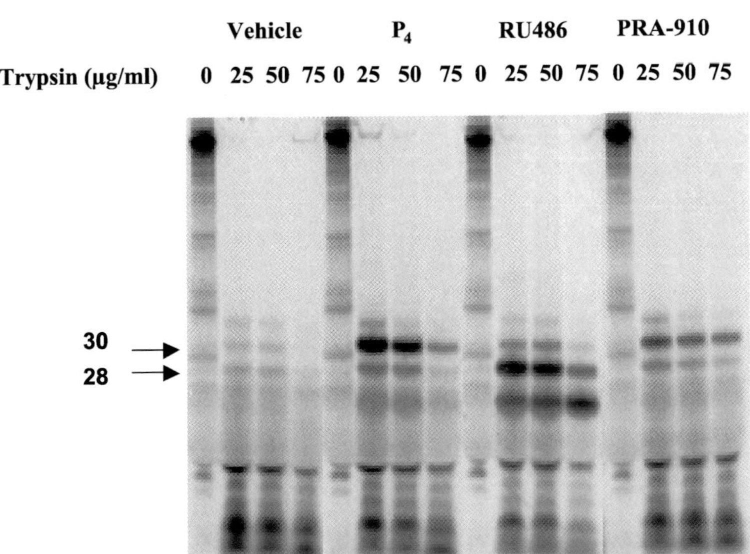

Fig. 4. The protease digestion pattern of PRA-910-bound PR suggests an agonist conformation is induced. The indicated ligands, or DMSO vehicle, were pre-incubated with the in vitro translated [^{35}S]-PR and the indicated amount of trypsin before analysis by sodium dodecyl sulfate polyacrylamide gel electrophoresis (SDS-PAGE) and autoradiography. The 30-kDa band is characteristic of agonist conformation and the 28-kDa band is characteristic of an antagonist conformation

not show any PR agonist activity at a dose up to 10 mg/kg, the highest dose tested (data not shown).

The second model used to assess PR activity in the rat was the uterine complement C3 assay (Lundeen et al. 2001). In the rat, estrogens induce uterine epithelial expression of complement component C3. This upregulation is suppressed when the estrogen is coadministered with a progestin and can be reversed by a PR antagonist (Fig. 6a). As shown in Fig. 6b, PRA-910 also antagonized the P$_4$ effect in this model. The mean ED$_{50}$ (dose that blocks the anti-estrogenic effect of progesterone on uterine C3 mRNA by 50%) in this model was 1.3+/–0.1 mg/kg for PRA-910 and 1.7 for RU-486. Treatment with 10 mg/kg of PRA-910

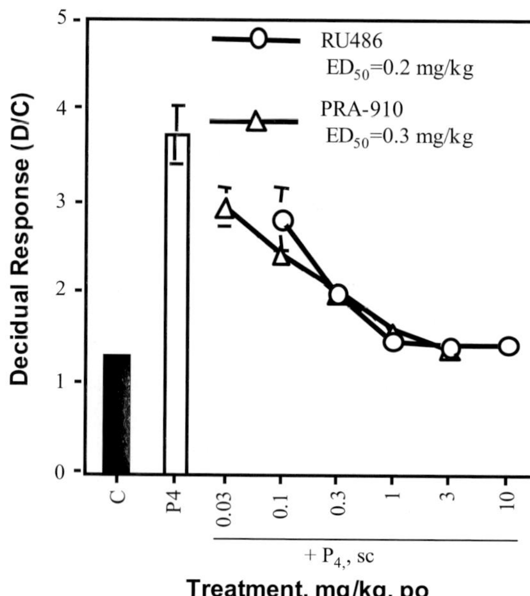

Fig. 5. The rat decidualization assay shows PRA-910 is a potent and efficacious PR antagonist in this species. Decidual response is measured as a ratio between the weight of the decidual horn (*D*) and that of the control, non-decidualized horn (*C*) in the same animal. Dosing is described in the materials and methods. Neither RU-486 nor PRA-910 had any agonist effect when treated alone up to 10 mg/kg (not shown)

alone (agonist mode) had no effect on C3 mRNA levels. Interestingly, both PRA-910 and RU-486 cotreatment with EE resulted in C3 mRNA levels above that observed for EE alone.

3.6   PR Agonist Activity of PRA-910 in the Monkey

A preliminary experiment was conducted to determine if PRA-910 acts as a PR antagonist by inducing menses in artificial cycling rhesus macaques. Treatment with PRA-910 did not induce early menses, suggesting this compound did not act as a PR antagonist in macaques. Fur-

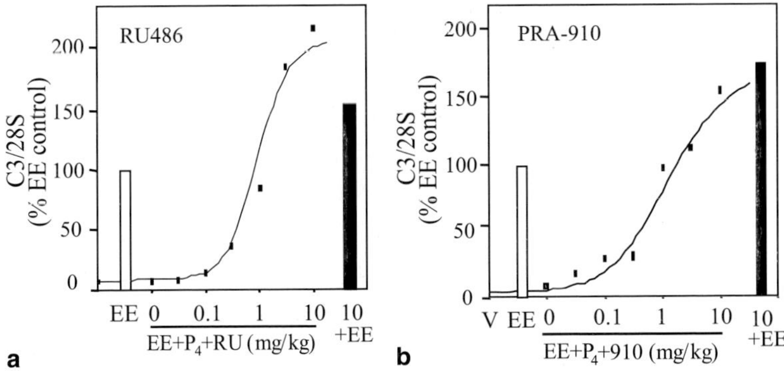

Fig. 6a, b. Effect of RU-486 (**a**) and PRA-910 (**b**) on rat uterine C3 mRNA levels. Vehicle (*V*) and ethinyl estradiol (*EE*) alone are shown along with EE+P_4 with increasing concentrations of either RU-486 (**a**) or PRA-910 (**b**). The *last bar* in each plot is the compound run at 10 mg/kg in the absence of P_4 (agonist mode). Data are determined as the ratio C3 mRNA/28S rRNA with the value of EE set at 100%

thermore, menses did not occur after P_4 withdrawal, indicating the compound may act as a PR agonist.

The activity of PRA-910 in the endometrium and oviduct in the non-human primate was evaluated using an established rhesus model (see materials and methods). In this model, 28-day treatment of estrogen-primed monkeys with 5 mg/kg PRA-910 resulted in extensive glandular sacculation (Fig. 7b, d) especially in the basalis zone, similar to what is seen with P_4 treatment. There was also evidence of stromal expansion (compare Fig. 7e and f), but endometrial mass and thickness were not significantly different from E_2-treated animals. This minimal effect on thickness may be attributed to the short 7-day treatment. Treatment with PRA-910, like P_4, inhibited epithelial cell proliferation, resulting in no identifiable mitotic cells in the glandular epithelial layer (Fig. 7d). Further analysis of cell proliferation is shown in Fig. 8, where PRA-910 inhibited both expression of Ki-67 and Brd(U) incorporation in the functionalis zone glands (Fig. 8b and f as compared to a and e). In contrast, PRA-910 did stimulate cell proliferation in the basalis zone where cell proliferation is induced by progestins, leading to a Brd(U) labeling

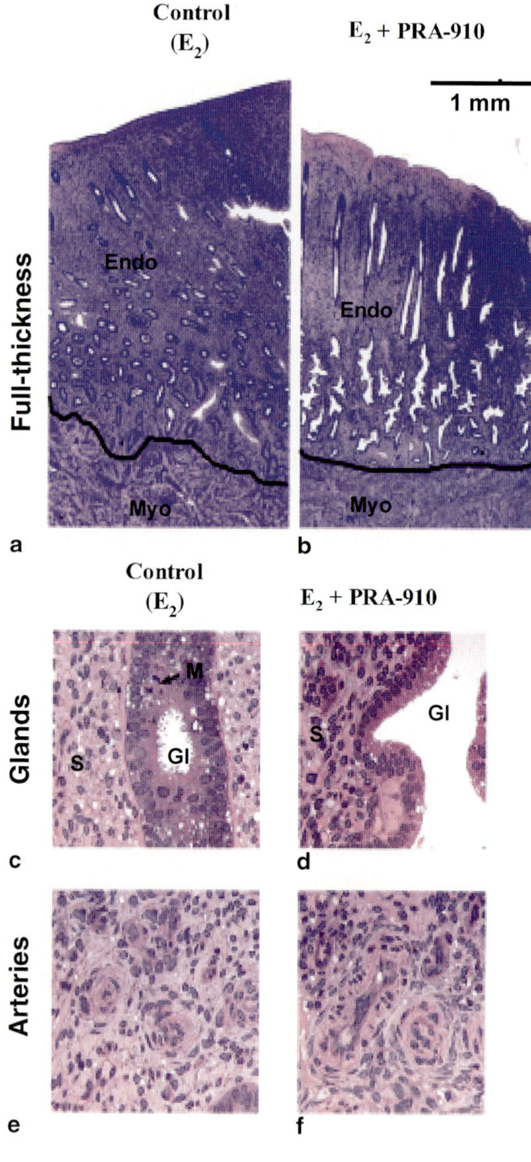

Control
(E₂)
E₂ + PRA-910
1 mm
Full-thickness
Endo
Endo
Myo
Myo
a
b
Control
(E₂)
E₂ + PRA-910
Glands
M
S
Gl
Gl
S
c
d
Arteries
e
f

Fig. 7a–f. Photomicrographs of GMA-embedded, hematoxylin-stained sections of the rhesus endometrium treated with E2 (control) or E2+PRA-910. Full-thickness endometrium is shown in **a** and **b**, and the *dark line* indicates the endometrial (*Endo*)/myometrial (*Myo*) border. Glands (**c** and **d**) and arteries (**e** and **f**) were photographed at 250×. *E2,* 17b-estradiol; *S,* stroma; *Gl,* gland; *M,* mitotic cells

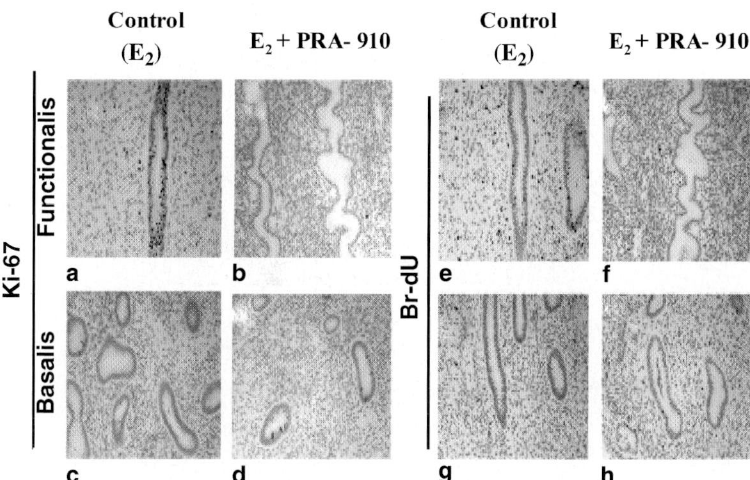

Fig. 8a–h. Photomicrographs of rhesus endometrium immunostained (*dark brown*) for Ki-67 (**a–d**) or Br-dU (**e–h**) from animals treated with either E2 alone (control) or E2+PRA-910. Both the basalis (**c, d, g, h**) and functionalis (**a, b, e, f**) zones of the endometrium are shown

index of 29 cells/1,000 glandular epithelial cells (Fig. 8d and h), similar to that observed after E_2+P_4 treatment (Slayden et al. 2006). Further indication of progestin-like activity of PRA-910 was determined by evaluation of ERα and PR expression. Figure 9 clearly shows the inhibition of both ER and PR expression in the glandular epithelium following PRA-910 treatment, similar to what is seen with P_4 treatment (Slayden et al. 2006).

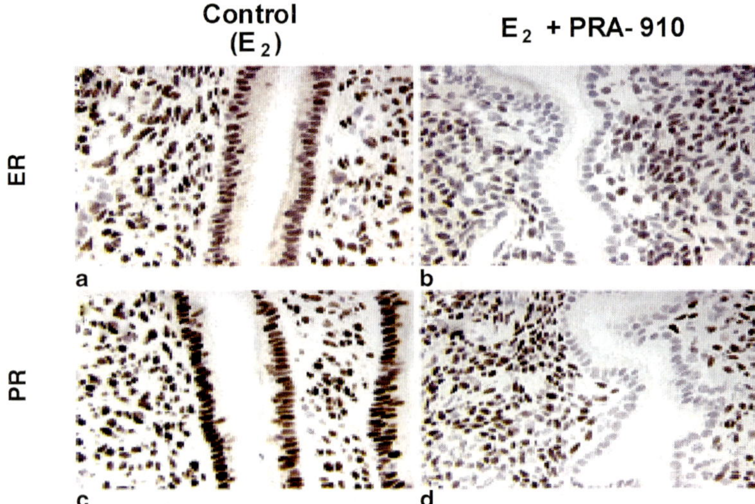

Fig. 9a–d. Photomicrographs of rhesus endometrium immunostained (*dark brown*) for estrogen receptor, ERα (**a** and **b**), and progesterone receptor, PR (**c** and **d**), in animals treated with E2 (control) or E2+PRA-910 as described in the materials and methods

Fig. 10a, b. In vivo steroid receptor selectivity of PRA-910. **a** The ability of PRA-910 to induce thymic involution was compared to the reference glucocorticoid dexamethasone (DEX). PRA-910, when dosed alone at 3, 10, and 30 mg/kg, was not significantly different (*) from vehicle control (*C*), indicating lack of glucocorticoid receptor (*GR*) agonism. Similarly, when dosed in conjunction with DEX, PRA-910 failed to reverse the reduction in thymic weight, unlike RU-486 (*RU*), a known GR antagonist. **b** PRA-910 was tested for androgen receptor (*AR*) antagonism by measuring the effect of treatment in conjunction with testosterone propionate (*T.P.*). TP markedly increases ventral prostate weight, which is not reversed by cotreatment with PRA-910 at 3, 10, and 30 mg/kg. 4-Hydroxy-flutamide (FL), a known AR antagonist, does reverse the effect of TP on ventral prostate weight. The *asterisk* indicates significant difference ($p<0.05$) from TP alone

3.7   Other Steroid Receptor Activity of PRA-910 in the Rat

PRA-910 was evaluated for both GR and AR activity in the rat at 3, 10, and 30 mg/kg. In adrenalectomized male rats, PRA-910 did not show significant glucocorticoid or anti-glucocorticoid activity at all doses tested (Fig. 10a) as measured by evaluation of thymus gland weights following 5 days of treatment. Evaluation of androgenic and anti-androgenic activity of PRA-910 in the rat was also evaluated. Following a 10-

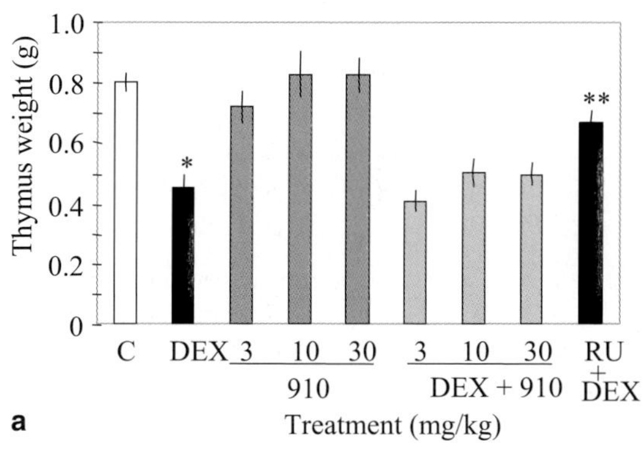

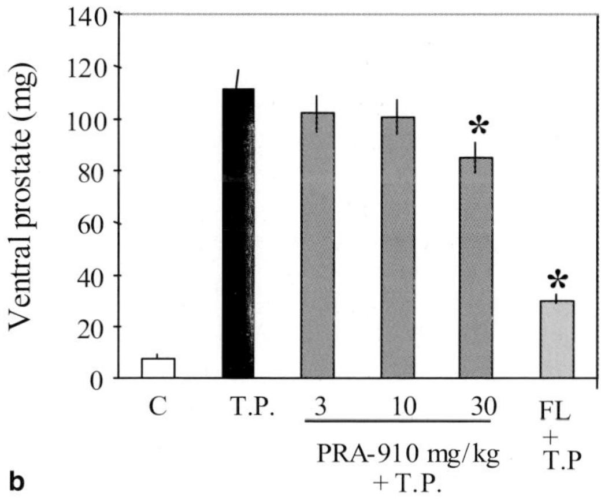

day treatment paradigm, PRA-910 did not show any androgenic activity in castrated male rats and showed weak but significant anti-androgenic activity at 30 mg/kg but not at lower doses (Fig. 10b).

4   Discussion

In this report, we describe the biological characterization of a novel, nonsteroidal PR modulator, PRA-910. PRA-910 was identified as a PR antagonist based on its in vitro activity on a reporter gene in COS-7 cells. The IC_{50} value of 20 nM in this assay correlates well with the binding affinity of 4.4 nM measured with [^{3}H]-labeled PRA-910 using T47D cell extracts. In addition to the human PR, PRA-910 binds with high affinity to the rat, rabbit, and monkey PR as well. Evaluation of PRA-910 in the T47D alkaline phosphatase assay resulted in a biphasic curve. At low doses (<100 nM), PRA-910 antagonized the progesterone-induced rise in alkaline phosphatase activity but with an average efficacy of 43%. However, at higher concentrations PRA-910 predominately showed PR agonist activity, inducing alkaline phosphatase activity with an EC_{50} value of approx. 700 nM. In the mammalian two-hybrid assay measuring the interaction between PR and SRC-3, PRA-910 also demonstrated potent PR antagonist activity (IC_{50}=21 nM, 70% inhibition) with weak, but fully efficacious, agonist activity at high concentrations (EC_{50} in the micromolar range). Interestingly, in a limited proteolytic digestion analysis with the human PR, PRA-910 produced a PR peptide pattern that was very similar to that of P_4 or other steroidal progestin-treated samples. In vitro, therefore, both the agonist and antagonist properties of PRA-910 can be measured depending on the concentration, cell type, and endpoint.

In vivo, the effects of PRA-910 depend on species. In the rat, very potent PR antagonist activity was observed, similar to that of RU-486. It thus appears that despite an approximately 10- to 50-fold difference in binding affinities and in vitro functional responses, the activity on the PR in vivo is similar between these two compounds. This effect was consistent at doses up to 10 mg/kg, the highest dose tested in the rat. However, when PRA-910 was studied in rhesus macaques, the compound switched its functional profile and behaved as a PR agonist at

5 mg/kg. Here we find PRA-910 blocks estrogen-induced epithelial cell differentiation in the oviduct, and importantly, induces glandular differentiation and suppresses DNA synthesis in the endometrial epithelium in estrogen-primed animals, two well-defined endpoints of progestational activity (Slayden and Brenner 1994). The selectivity of this compound for the PR as well as the selectivity measured in vivo would preclude nonselective activity as a cause for this species-specific profile switch.

It is known that weak agonists can function as antagonists at concentrations below full efficacy where the molecule can still bind and compete with the more potent agonist present. For example, estriol, a weak ER agonist will exhibit partial antagonism of estradiol at concentrations below the full efficacy for estriol (Melamed et al. 1997). Another nonsteroidal PR antagonist, RWJ47628, also shows agonist activity at higher concentrations (Tabata et al. 2003). However, neither of these compounds is known to exhibit opposing species-selective activities or show the diversity of in vitro responses as does PRA-910.

The mechanism by which PRA-910 elicits this opposing biology in different cell contexts and species remains to be elucidated. Several possibilities exist. First, the answer may lie in the sequence of the PR protein itself. For example, in the rat, PRA-910 may induce an antagonist conformation preventing coactivator recruitment while the conformation generated in the primate permits such protein interactions necessary for transcription and agonist activity. Preliminary data, however, suggest that PRA-910 behaves similarly in a cotransfection reporter assay using a rat PR construct (J. Bretz and M.R. Yudt, unpublished data). Ultimately it will be of interest to determine if PRA-910 induces unique species-specific conformations, and how these conformations could drive the observed biology. Beyond potential species-specific receptor conformations, how PRA-910 liganded PR interacts with coactivators/corepressors in each cell type and species remains unknown. It is possible that different ratios of coactivators and corepressors in these various cell types and species contribute to balance of agonist or antagonist activity conferred by PRA-910. Furthermore, examination of gene expression patterns by PRA-910 in different cell types and species may help us understand the PR agonistic and antagonistic effect of this compound. Studies are currently underway to address these questions to and further understand the unique biological profile of PRA-910.

Acknowledgements. The authors wish to thank Dr. S.G. Lundeen for numerous contributions to this project. Dr. O. Slayden is a recipient of grant #RR00163 from Oregon National Primate Center.

References

Allan GF, Leng X, Tsai SY, Weigel NL, Edwards DP, Tsai MJ, O'Malley BW (1992) Hormone and antihormone induce distinct conformational changes which are central to steroid receptor activation. J Biol Chem 267:19513–19520

Beck CA, Estes PA, Bona BJ, Murocacho CA, Nordeen SK, Edwards DP (1993) The steroid antagonist Ru486 exerts different effects on the glucocorticoid and progesterone receptors. Endocrinology 133:728–740

Chwalisz K, Perez MC, Demanno D, Winkel C, Schubert G, Elger W (2005) Selective progesterone receptor modulator development and use in the treatment of leiomyomata and endometriosis. [erratum appears in Endocr Rev 2005 Aug;26(5):703]. Endocr Rev 26:423–438

Foidart JM (2000) The contraceptive profile of a new oral contraceptive with antimineralocorticoid and antiandrogenic effects. Eur J Contracept Reprod Health Care 5 [Suppl 3]:25–33

Lundeen SG, Zhang Z, Zhu Y, Carver JM, Winneker RC (2001) Rat uterine complement C3 expression as a model for progesterone receptor modulators: characterization of the new progestin trimegestone. J Steroid Biochem Mol Biol 78:137–143

Lydon JP, DeMayo FJ, Funk CR, Mani SK, Hughes AR, Montgomery CA Jr, Shyamala G, Conneely OM, O'Malley BW (1995) Mice lacking progesterone receptor exhibit pleiotropic reproductive abnormalities. Genes Dev 9:2266–2278

McDonnell DP, Clemm DL, Hermann T, Goldman ME, Pike JW (1995) Analysis of estrogen receptor function in vitro reveals three distinct classes of antiestrogens. Mol Endocrinol 9:659–669

Melamed M, Castano E, Notides AC, Sasson S (1997) Molecular and kinetic basis for the mixed agonist/antagonist activity of estriol. Mol Endocrinol 11:1868–1878

Psychoyos A (1973) Hormonal control of ovoimplantation. Vitam Horm 31:201–256

Slayden OD, Chwalisz K, Brenner RM (2001a) Reversible suppression of menstruation with progesterone antagonists in rhesus macaques. Hum Reprod 16:1562–1574

Slayden OD, Brenner RM (1994) RU 486 action after estrogen priming in the endometrium and oviducts of rhesus monkeys (Macaca mulatta). J Clin Endocrinol Metab 78:440–448

Slayden OD, Koji T, Brenner RM (1995) Microwave stabilization enhances immunocytochemical detection of estrogen receptor in frozen sections of macaque oviduct. Endocrinology 136:4012–4021

Slayden OD, Nayak NR, Burton KA, Chwalisz K, Cameron ST, Critchley HO, Baird DT, Brenner RM (2001b) Progesterone antagonists increase androgen receptor expression in the rhesus macaque and human endometrium. J Clin Endocrinol Metab 86:2668–2679

Slayden OD, Zelinski MB, Chwalisz K, Hess-Stumpp H, Brenner RM (2006) Chronic progesterone antagonist-estradiol therapy suppresses breakthrough bleeding and endometrial proliferation in a menopausal macaque model. Hum Reprod 21:3081–3090

Tabata Y, Iizuka Y, Shinei R, i Kurihara K, Okonogi T, Hoshiko S, Kurata Y (2003) CP8668, a novel orally active nonsteroidal progesterone receptor modulator with tetrahydrobenzindolone skeleton. [erratum appears in Eur J Pharmacol 2003 Apr 4;465(3):299]. Eur J Pharmacol 461:73–78

Zhang Z, Funk C, Glasser SR, Mulholland J (1994) Progesterone regulation of heparin-binding epidermal growth factor-like growth factor gene expression during sensitization and decidualization in the rat uterus: effects of the antiprogestin, ZK 98.299. Endocrinology 135:1256–1263

Zhang Z, Lundeen SG, Zhu Y, Carver JM, Winneker RC (2000) In vitro characterization of trimegestone: a new potent and selective progestin. Steroids 65:637–643

Zhang P, Terefenko EA, Fensome A, Wrobel J, Winneker R, Lundeen S, Marschke KB, Zhang Z (2002a) 6-Aryl-1,4-dihydro-benzo[d][1,3]oxazin-2-ones: a novel class of potent, selective, and orally active nonsteroidal progesterone receptor antagonists. J Med Chem 45:4379–4382

Zhang P, Terefenko EA, Fensome A, Zhang Z, Zhu Y, Cohen J, Winneker R, Wrobel J, Yardley J (2002b) Potent nonsteroidal progesterone receptor agonists: synthesis and SAR study of 6-aryl benzoxazines. Bioorg Med Chem Lett 12:787–790

Ernst Schering Foundation Symposium Proceedings

Editors: Günter Stock
Monika Lessl

Vol. 2006/1: Tissue-Specific Estrogen Action
Editors: K.S. Korach, T. Wintermantel

Vol. 2006/2: GPCRs: From Deorphanization
to Lead Structure Identification
Editors: H.R. Bourne, R. Horuk, J. Kuhnke, H. Michel

Vol. 2006/3: New Avenues to Efficient Chemical Synthesis
Editors: P.H. Seeberger, T. Blume

Vol. 2006/4: Immunotherapy in 2020
Editors: A. Radbruch, H.-D. Volk, K. Asadullah, W.-D. Doecke

Vol. 2006/5: Cancer Stem Cells
Editors: O.D. Wiestler, B. Haendler, D. Mumberg

Vol. 2007/1: Progestins and the Mammary Gland
Editors: O. Conneely, C. Otto

Vol. 2007/2: Organocatalysis
Editors: M.T. Reetz, B. List, S. Jaroch, H. Weinmann

Vol. 2007/3: Sparking Signals
Editors: G. Baier, B. Schraven, U. Zügel, A. von Bonin